Betting the Farm on a Drought

Betting the Farm on a Drought

Stories from the Front Lines of Climate Change

Seamus McGraw

University of Texas Press
Austin

Support for this book comes from an endowment for environmental studies made possible by generous contributions from Richard C. Bartlett, Susan Aspinall Block, and the National Endowment for the Humanities.

Printed in the United States of America
First edition, 2015

♾ The paper used in this book meets the minimum requirements of ANSI/NISO Z39.48-1992 (R1997) (Permanence of Paper).

LIBRARY OF CONGRESS CATALOGING-IN-PUBLICATION DATA
McGraw, Seamus, author.
Betting the farm on a drought : stories from the front lines of climate change / by Seamus McGraw. — First edition.
pages cm
ISBN 978-0-292-75661-8 (hardback)
1. Climatic changes—United States—Popular works. I. Title.
QC903.2.U6M34 2015
363.738′740973—dc23
2014027613

doi:10.7560/756618

For my old man

Contents

Betting the Farm on a Drought

My son, Liam, at the drill pad.

{ ONE }

Sundance

It had been weeks since I had been there, and now, as my young son, Liam, and I rounded the corner of Ellsworth Hill Road near the edge of my family farm in my rattletrap old Chevy Blazer, it loomed into view, a massive cloud of dust and sand and diesel exhaust billowing up into the sky and all but blotting out the moon. Backlit by what must have been a hundred high-powered lights atop the seventy-foot-tall, hundred-yard-square Mayan pyramid the drillers had carved into our hillside, it looked as if some sort of bizarre ritual—a sacrifice, perhaps—was playing out in front of us.

Maybe it was.

Two years earlier, Liam had been with me when we walked that ground with the surveyors for Chesapeake Appalachia, pacing out what they thought would be the perfect place to site the first of what they anticipated would be six mile-and-a-half-deep natural gas wells that would then snake out a mile and a half to suck out as much gas as they could from the Marcellus Shale beneath our little corner of northeastern Pennsylvania. The lead surveyor, a tall man with a deep-fried Texas accent, had tried his best to persuade my son to drive in the first stake, but when he offered the six-year-old boy the four-pound mallet, Liam hightailed it behind a blackberry bramble. A look of confusion and maybe a little bit of fear etched its way across my son's brow.

I glanced in the rearview mirror as I pulled into the long dirt driveway. I could see that look cross his face again. And then I glanced at my own reflection. The look was carved onto my brow, too.

Staring at this massive, churning industrial operation in what used to be my family's pasture, the place where as a boy I chased Angus cross cows on horseback pretending to be a cowboy, I listened to the otherworldly roar of the diesel engine that powered it all, the earth literally pulsing beneath me. I was surprised by my own feelings.

I'd thought I was prepared for it. I had seen this operation before—the military precision of the small army of roughnecks and technicians as they revved up the intimidating array of high-tech equipment, including the forty-foot-long battleship of a machine that would blast a toxic cocktail of water and sand containing up to a dozen chemicals, some of them kept secret even from the landowner, deep into the earth at more than 9,400 pounds of pressure per square inch to pry open tiny fissures in the rock and release the natural gas trapped inside it.

But now that it was happening here, on my land, I could understand my own son's wordless confusion. I could appreciate in a visceral way, maybe for the first time, why the word to describe this process—fracking—stirs such fear. I could even feel the stirring of that fear myself.

Truth be told, that sense of unease had been gnawing at me for a while, ever since we had first signed the contract to allow the drillers onto our land. And while it was still unformed, it had taken on a kind of urgency the summer before, when I started out on the road to promote my last book, *The End of Country*, which detailed the difficult issues my family and my neighbors faced as we grappled with the question of whether to risk all we had and allow the drillers to bore their way into our lives.

In town after town, at college after college, every place I was invited to speak, I saw the deep fractures that this process had exacerbated. It was as polarized and polarizing an issue as any I've ever encountered. But more than that, I came to see it as a kind of metaphor. It was, in miniature, a portrait of all the deep divisions that snake through our political and social discourse. It is no exaggeration to say that when talking to the vast majority of people I had met on the road—and there were a couple of thousand of them—if I could tell where they stood on the issue of fracking, I could tell with an alarming degree of accuracy where they stood on a half dozen or so other contentious issues, from abortion and gun rights to the growing gulf between the haves and the have-nots in this country. Nowhere was that rift more jagged than on the question of the risks and possible responses to the changing climate, an issue that may be the most consequential of our time.

It is, of course, perfectly understandable that people tend to cluster in almost tribal groups when faced with issues that are so multifaceted and so complex, so full of nuance and uncertainty, every element of them fraught with both promise and peril. Hell, even scientists who spend their entire professional lives studying the risks associated with the changing climate admit that they sometimes have a hard time wrapping their minds around the magnitude of it all.

After all, insulated as we are, it's no surprise that most harried Americans find little time to ponder the complex network that links their consumption of everything to the rising sea levels that threaten to erase from the map some South Pacific island nation they've never heard of before. To a nation that for the most part knows polar bears only as furry computer-generated pitchmen who turn up on television between Thanksgiving and New Year's to sell them Coca-Cola, the whole idea of melting ice caps is, to borrow a phrase used by New Jersey governor Chris Christie in the wake of Hurricane Sandy, a little "esoteric."

Americans, who are by the millions treading water just trying to keep pace with their mortgages and their rising grocery bills, would rather not think about the issue at all. And if they are prodded to take a position, they'll likely take the one espoused by the coven of talking heads they usually trust, be it Fox News or MSNBC.

Nor is it any surprise that in the chaotic maelstrom that is modern media, only the most strident voices are heard above the din, and so when we manage to find any discussion about this at all, the few words we catch are the most divisive.

The tragedy of it is that across the whole cultural landscape of this country, all of those factors are conspiring to make the fractures deeper and wider. It has become almost clichéd to say that the country is now more divided than it has been at any time since perhaps the Civil War, but the evidence of that, and the consequences of it, becomes clearer with each passing year.

There's a line I often use when I'm speaking: "If the melting ice caps and the rising oceans don't get us, we're all going to drown in the viscera of each other's gored oxen."

It usually elicits a rueful chuckle from the audience. Sometimes you can get a laugh just by stating the obvious.

But there is a part of me—call it naive, if you like—that can't help but believe that somewhere out there in America, all but drowned out by the din, is a native strain of reason that could be cultivated, a peculiarly American kind of common sense and courage that would give us the strength to at least begin an honest discussion about the challenges we face.

I don't know. Maybe that's what I was looking for when I drove to the farm that day with Liam.

Maybe I was looking for it in the contours of the changing land itself. Maybe I was looking for it in my son's expression. Maybe I was looking for it in myself.

It had not been an easy decision to let the drillers onto our land a few years earlier. Not for me. Not for my family. Not for our neighbors, most of them former dairy farmers who had been on this land for generations and who, one by one, had eventually been forced out of business, strangled almost to death by the twin tendrils of bad farm policy and spiraling energy prices. To us, this land was more than the sweet spot on some U.S. Geological Survey map of the gas-rich Devonian shales that underlie most of northern Appalachia. It was home. The sum of who we were. The place where I used to chase cows on horseback.

My family had been there for more than four decades. What

started out as a weekend retreat quickly became an obsession, and before long it became a full-time job as my father indulged his dream of becoming a gentleman farmer and my mother chased her dream of becoming a character in one of those frontier romance novels—buckskin bodice rippers, she called them—that she so adored. Mom managed to achieve her dream; Dad, not so much, and eventually, he quit trying.

As for me, when I turned eighteen I shook the dust of that place off my boots, headed away to college for a while, failed at that, and then failed at a series of jobs and marriages until I accidentally stumbled into journalism and never figured out how to drift back out of it. And yet the place was always with me. It defined me. I went back every chance I got. My sister was married there. So was I, the third time. My father died there.

It was there, at my father's deathbed, that I learned fractures can be deepened and widened, but they can also be closed. I learned it in his room with him just two nights before he finally succumbed to the pancreatic cancer that, unbeknownst to either of us, had already been eating away at him just months earlier when he and I had silently worked up a sweat while building a bluestone patio beside the barn for my wedding. I had been at work, three and a half hours away, at a newspaper in New Jersey—my sister and I would work all day and then drive to the farm each evening to maintain a vigil before turning the hospice duties back over to my mother—when my mother called me, frantic and out of breath.

"Your father wants to go down and sit on the porch," she said, her voice cracking. "He's seventy pounds. I can't do it; I can't carry him. I'll kill him. But he's yelling at me."

I tried to calm her down, and told her I was on my way. I made it to the farm in record time. And when I walked into the room, he was, as my mother had warned me, on a tear. I looked at him. Lying there, in that hospice bed, he looked so small and frail, and yet he was full of rage and, I have no doubt, fear. A bit of fear infected me as well.

Things hadn't always been easy between my father and me.

Maybe the pieces of us that were different were just too different. Maybe the pieces of us that were similar were just too similar. We were both proud and headstrong, and had certainly both made our share of mistakes, and there were long periods during my younger years when the disappointment we felt toward each other got the best of us. We once went for two years without speaking to each other. But time and age have a way of changing your expectations, and eventually my father and I rediscovered each other and settled into a rhythm. We stopped wishing the other would be something else and came to respect and love each other for who we were, or at least, for who it was that we wanted to be. And for the rest of my father's life, we maintained—no, we nurtured—that relationship, talking, when we talked, about those things we shared and carefully tending the silences when necessary. Or at least we did until the day my mother called.

"I want to go downstairs," my father demanded, his voice surprisingly strong for a man in his condition.

"You can't," I said firmly. "You're too weak. She couldn't carry you," I said, pointing to my mother, who was now sobbing in a corner of the room. "And if I try to carry you down, there's a good chance I'll kill you."

"Goddamn it," my father croaked back. "Take me downstairs!"

"No," I said, just as firmly.

My father reached up with a skeletal hand and grabbed the side railing of the hospice bed and began violently tugging at it. "Listen, you little son of a bitch," he sputtered, "take me downstairs or I'll rip this goddamned thing off and wrap it around your neck!"

"Dad!" I said, with as much forced joy as I could muster. "Where have you been? I've missed you!"

The old man stopped and cocked an eyebrow at me. And then a smile crossed his face. It was the last time I ever saw my father laugh.

That moment, like so many others from the time I first set foot on the land, was frozen forever for me, part of a place that I would always picture in my mind's eye as being the way I remembered it. But I knew it wasn't that way anymore.

As the dairy farms that once surrounded us failed, those who could leave did, selling their acreage, often in small chunks, to people from New York and New Jersey who imported with them a fantasy of country living. Little by little, the whole way of life I had known as a boy was vanishing. It was, as one of my neighbors put it, "the end of country."

And now, the drillers were here.

The way I saw it then, and the way I still see it, there was a sense of inevitability to it all. It wasn't just about the money. Although some of us, like my own family, were offered hundreds of thousands of dollars to lease our land, with the promise of perhaps millions more when and if the wells came in, others, like my neighbor across the road who signed with a smooth-talking land man who turned up before the full potential of the Marcellus was really understood, got a pittance, just enough to pay their property taxes.

No, it was about something more important. The way we saw it, maybe the gas in the Marcellus could buy us one more chance.

Sure, there were risks. They were real and they were profound. There was the risk that fracking fluid could either spill on the surface or—though most analysts agree it's far less likely—migrate up from deep underground to possibly contaminate groundwater and aquifers. There was the risk that methane, the key component of natural gas, could waft up into drinking-water supplies as a result of carelessness, recklessness, bad cement, or just bad luck. In fact, that had happened, repeatedly—most famously just a few miles away in the little village of Dimock, Pennsylvania, a place that has become in many respects, ground zero for the debate over fracking. There was the threat that the whole thirsty process—it takes millions of gallons of water to break up the shale, and at least 30 percent of it remains underground, lost to the water cycle forever—would tax water supplies. There were risks associated with disposing of the water, not just the water that flowed back from the fracking operation but the even more noxious witch's brew that the earth itself cooks up in the shale, a slightly radioactive, highly saline and heavy-metal-laden water that flows up to the surface for the lifetime of the well, still seeping up decades hence, when not as many people are watching as closely. Some of that water was treated and recycled. Some was treated and released. Some was pumped into deep-injection wells, many of them in Ohio, in a process that on rare occasions had triggered earthquakes like the series of small tremors, up to magnitude 4.0, that rattled Youngstown, Ohio, in 2011 after deep-well injection operations began there, forcing state officials to shut down the facility. There was also the risk to the air: methane leaking from wellheads, and pipelines, and compressor stations, and the steady noxious cloud of diesel fumes produced at virtually every step of the drilling process. For all the grand talk about how much better for the environment natural gas is than oil as a fuel, drillers still rely heavily on good old-fashioned diesel to get it.

Those risks, the industry and its boosters insisted, were mechanical, and were therefore manageable.

But there were other risks that were far less easy to predict, and far more difficult to contain.

When the drillers showed up, so, too, did the advocates on both sides of the issue—movie stars, activists, industry shills, and self-promoting documentarians. In much the same way that the process of fracking injects astounding amounts of pressure into the deeply buried rock to exploit existing fractures, well-funded pseudo-grassroots organizations blasted their people and resources into these already wounded communities to exploit the social fractures that had long divided them.

Nowhere was that more evident than on Carter Road, a little dirt track that snaked behind a hillside in Dimock, a place where fifteen families had found their water tainted by methane after drilling had begun nearby. The Pennsylvania Department of Environmental Protection (DEP) had laid the blame squarely at the driller's feet. The driller was ordered to shut down operations in that area, and for a while the driller, Cabot, had provided water to the affected residents. A lawsuit filed by the families had forced the company to do so. But when that lawsuit was settled, and when the DEP decided to lift its ban on Cabot's operation, the water deliveries stopped.

When they did, the battle lines were drawn. Prodded by activists who argued that the water was contaminated by more than just methane, the federal Environmental Protection Agency stepped in, but when the agency could find no evidence of significant contamination beyond what the DEP had found, it stepped aside. In its wake, the activists on both sides stepped in, lining up neatly and orderly on their respective sides of the fractures in the community.

By late 2011, it was not at all unusual to drive down Carter Road, a place that ordinarily might see no more than an occasional rabbit cross the road, and find it choked with cars and trucks and buses and news vehicles, as a throng of true believers clustered around *Gasland* auteur Josh Fox and actor Mark Ruffalo in front of a house festooned with "Ban Fracking" signs, or one house over, where pro-fracking filmmaker Phelim McAleer held court in a yard where pro-fracking signs sprouted like dandelions on the lawn.

Lost in the uproar between the two sides, however, was any real discussion of the potential benefits that might balance those risks and increase the incentive to manage them. And those benefits, if they could be realized, would accrue not just to the few landowners lucky enough to squeeze a few bucks out of the drillers. There was the potential of more far-reaching benefits.

The truth was, at precisely the same time that the fractures on Carter Road were deepening, so, too, was the crisis of climate in much of the nation. The country was poised to experience one of its hottest years on record, a year that would bring with it raging wildfires in the West, a drought that very nearly rivaled the one that sparked the Dust Bowl in the Midwest and Southwest, and ultimately, the murderous fury that was Hurricane Sandy when it slammed into the East Coast the following September.

And yet, while the majority of scientists—97.1 percent of those who had written about the subject, according to a 2013 study by

John Cook of the University of Queensland Climate Change Institute[1]—were convinced that human activity, particularly fossil fuel consumption, was significantly responsible for the rising temperature and the erratic weather conditions that accompanied it, the American people remained deeply divided. So politically fraught was the debate that in 2012, North Carolina's conservative legislature passed a law prohibiting the state from using scientific predictions of sea level rise as a basis for developing coastal policies.

But despite those deep cultural divisions, despite the deep ambivalence of many Americans, despite the all-or-nothing approach of environmental activists on the one hand and pro-drilling activists on the other, despite the partisan-induced paralysis among lawmakers on both the state and federal levels, the United States had, in the four years leading up to those cultural clashes on Carter Road, measurably reduced its carbon output, which by 2012 dropped to the lowest level in nearly twenty years.[2]

There were a couple of reasons for that. Increased fuel efficiency was one of them, analysts said. So was the sluggish economy. It's a sad truth that if you want to reduce your carbon output, choking your gross domestic product to death will do it, and between 2007, when the economy fell off a cliff, and 2013, we pretty much did just that, creeping along at an anemic growth rate that averaged between 2 and 3 percent, keeping demand for energy more or less flat, according to figures released by the International Energy Agency (IEA).

To be sure, the growth of renewables played a critical role as well. And while they still produce a comparatively small fraction of the energy we consume, wind power increased its share of the market in the United States by 26 percent in 2012 and was expected to increase another 17 percent in 2013, and solar energy nearly doubled, according to the U.S. Energy Information Administration. Overall, the EIA projected that all non-hydropower renewables would increase their market share of our total energy consumption by about 5 percent the following year.

"Just to put that in perspective," Laszlo Varro, head of the International Energy Agency's gas, coal, and power market division, told me when I called him at a disturbingly late hour and found him still in his office in Paris, "the increase in renewable energy in the United States in the past six or seven years is around one and a half times larger than the [total] wind and solar production of Germany," a country widely considered to be a world leader in renewable energy.

But all of those factors pale in comparison to the one key element that has contributed to the decline in U.S. carbon emissions, say Varro and other experts: natural gas, which, when burned, produces 50 percent as much carbon as coal and 30 percent less than oil, and in particular shale gas from deposits like the Marcellus.

"In the past five years there's little doubt that gas . . . was the single most important factor of bringing down carbon dioxide emissions," Varro said.

As drilling on the Marcellus and other shales has expanded exponentially in the past several years, driving down prices, gas has increasingly displaced coal as the fuel of choice for generation of electricity. The amount of energy in the United States generated by coal has decreased by 25 percent, while gas has increased its market share by 50 percent. In fact, Varro said, coal consumption in the American electrical energy sector has decreased by an amount equal to the entire electrical production of a "medium-sized European country, like England."

And yet, he and others acknowledge, as the world's geographic poles thaw, the political poles seem to have become bigger and frostier, with Pennsylvanians and all Americans becoming increasingly divided over the whole idea of fracking for natural gas and its benefits and risks.

That ambivalence among the public indicates how reports of the potential benefits of the natural gas boom have been tempered by deep concerns about the environmental consequences of the process used to extract the gas, and further stoked by what even the industry acknowledges is its history of seeming to be less than transparent with the public.

The ambivalent public attitude also reflects grave concerns

about the long-term impact of the amount of fossil fuel required simply to extract and process the natural gas, fuel that in many cases is still far dirtier than the gas itself, and about the release of methane during the process. While methane remains in the atmosphere a much shorter time than carbon dioxide does, it is a far more virulent greenhouse gas while it's up there.

There are also fears that increasing reliance on natural gas will undermine efforts to develop renewable forms of energy. The worry, expressed by many in the environmental community, is, as *New York Times* columnist Thomas Friedman wrote in 2012, that wholesale extraction of natural gas will "undermine new investments in wind, solar, nuclear and energy efficiency systems—which have zero emissions—and thus keep us addicted to fossil fuels for decades."

Those issues had certainly drawn the battle lines between two highly partisan and highly motivated camps that staked out their positions on Carter Road.

Looking for a little deeper insight, I sought out a guy who I knew could see across the fractures. As the former head of Pennsylvania's Department of Conservation and Natural Resources, John Quigley has a personal understanding of the motivations of people on both sides of the issue, insights that he has sharpened considerably in the years since, while traveling the country to lecture on the natural gas boom.

He often begins his talks by introducing himself with the words "I'm John Quigley and I'm from the state of hyperbole."

As he puts it, the voices that have come to dominate the debate are in danger of becoming as ossified by their ideology as the rock itself.

"On the left you have the moratorium crowd, the Alec Baldwins and the Josh Foxes of the world that want to demonize gas," Quigley told me. "'*The evil fossil fuel companies are out to kill us and rake in profits*,' and that's just . . . cartoon villainy. . . . And on the business side you have the old-boy network, the rough-and-ready, plucky entrepreneurs . . . who don't want to be told what to do. Scratch any of them and they're right-wingers."

The two sides, he said, are fast becoming so intractable that whatever advantages natural gas could offer in terms of reducing our reliance on dirtier, riskier fuel sources and American dependence on coal could easily be reversed. With the two sides locked in a stalemate, he said, there is a toxic political atmosphere, and one of the results is that little leadership has thus far emerged on a state or federal level to deal with either the environmental or the economic challenges presented by this development.

The reduction in carbon dioxide emissions that has come from switching electrical power plants from coal to gas, for example, has happened almost entirely as a result of market forces. The problem, said Quigley, is that "this reduction in CO_2 is illusory . . . because it's driven by Adam Smith's invisible hand and that's going to go away. It's temporary unless we take the bull by the horns and make something happen.

"There's a reduction in emissions, everybody feels good, but it's not a lock that this is going to last," he told me. There were, he argued, a number of reasons that emissions had declined. Warmer weather, for example, was one reason for it. Increased energy efficiency in everything from air conditioners to SUVs was another. "It's not just gas. Gas is probably one of the most significant things, but it was the combination that resulted in our emissions reductions, and anybody who thinks that they've reached the happy point where Adam Smith is going to lead us away from climate change, it isn't going to happen," Quigley said.

"We need for the first time in the history of this country a conscious energy policy that is more than aircraft carriers in the Persian Gulf," he said. The way he sees it, without a conscious national energy policy that takes advantage of natural gas and drives alternative energy on the back of gas, a policy that uses the political muscle of the gas industry to drive it all, we are "swinging and missing."

If the standoff on Carter Road demonstrated anything, it was the dire need for less-strident voices in the debate and the kind of consensus that could help build the political will to propel that kind of policy.

Increasingly, there are signs that less-strident voices are beginning to elbow their way into the public discourse. In August 2012, billionaire businessman, philanthropist, and then–New York City mayor Michael Bloomberg, along with the late George P. Mitchell, the Texas driller whose company launched the shale gas boom at the end of the last century and who has since passed on, wrote in the *Washington Post* that "fracking for natural gas can be as good for our environment as it is for our economy and our wallets, but only if done responsibly," while they criticized the industry for attempting to "gloss over" what they described as "legitimate concerns about its impact on water, air and climate."

Bloomberg followed that up with a $6 million grant to the Environmental Defense Fund, an organization that, according to its vice president of energy, Jim Marston, believes that "there are potentially large benefits to natural gas but only if done right, and in many places, [like] production, storage, transportation, natural gas is not being done right." According to a press release, the Bloomberg grant will be used to study ways to "minimize the environmental impacts of natural gas operations through hydraulic fracturing."

Among other things, some of that money will go toward funding a study to determine how much greenhouse effect–inducing methane is released into the atmosphere through the entire life cycle of natural gas production and consumption, Marston said.

Bloomberg's foray into the great debate included a call for increased government regulation on state and federal levels to improve five aspects of the operation: greater disclosure of the chemicals used in the process; tighter standards for well construction and operation; greater regulation of water consumption and disposal; plans to limit the negative impact of development on roads, communities, and the environment; and a greater focus on air pollution issues, with regard not just to fugitive methane but also to the amount of other fossil fuels used in the life cycle of natural gas.

And it comes on the heels of an exceedingly blunt statement at a speech in Houston in 2012 from IEA's executive director, Maria van der Hoeven, in which she argued that while natural gas expansion in the power sector "has caused emissions to fall rapidly," concerns about the risks are real, and the industry in many places around the world has been lax about acknowledging its failures and responding to them. If the industry continues to fail in that regard, she warned, "there is a very real possibility that public opposition to drilling for shale gas will stop the unconventional gas revolution and fracking in its tracks."

There are signs that at least some in the industry are beginning to hear the warnings coming from people like van der Hoeven and Quigley.

"What matters now is engagement," said Andrew Place, corporate director for energy and environmental policy at EQT, one of Pennsylvania's most active natural gas drillers, and a former deputy secretary of energy deployment at the DEP. "You need to get in a room with people . . . in the rational middle that are not throwing mortars at each other in both directions," he said.

To be sure, he said, "we cannot escape the fact that affordable energy is essential to our economy.

"Once you understand that, then you look at . . . all the pros and cons across the board, impacts, benefits, costs, what's the time line for implementing renewables and so on. Then you can have a robust discussion."

But the industry, he said, must also come to terms with the fact that it has to work to reduce its impact, and that, he said, is going to require a recognition that increased governmental oversight is probably inevitable. "You have to embrace regulation," he said.

That is precisely the message that the IEA has been trying to get across, Varro told me. "Overall we are optimistic about the . . . gas potential of the United States," he said. "If the U.S. shale revolution continues, then the U.S. . . . has the capacity to increase gas production at reasonable gas prices."

And even if Adam Smith's invisible hand pulls a card out of its sleeve and gas prices rise moderately, the cost of construction of new coal plants, costs that are certain to increase as a result of tougher standards for those plants, coupled with the regulatory uncertainty, make it probable that gas will continue to supplant coal as a fuel source for power generation. That, he said, means fewer greenhouse emissions, though it will almost certainly seesaw for a few years.

"But what if you have a political backlash, a ban on hydrofracking, let's say? In that case, the U.S. goes back to importing expensive energy from South America and the Middle East. In that case, U.S. gas prices go high again. And the U.S. will again face national security concerns."

The IEA, he said, has determined that the biggest noncommercial gas production will be in three countries: the United States, China, and Australia, which are also rich in coal.

Varro warned that if the public outcry against the excesses of the gas industry were to become strident and pervasive enough, if the still-nascent drive to establish bans on hydrofracking around the globe were successful, as much as 75 percent of that lost natural gas would be replaced not by other gas sources, but by coal.

"In that case, we lose the advantage," he said.

The IEA, he said, takes the position that "banning fracking and rolling back shale gas would be a really bad thing. It's somewhat ironic that environmentalists are not behind this. We also concluded that their concerns are absolutely legitimate. The usual natural gas industry talking point is that there's nothing to worry about, we are the experts, we know best. But we didn't buy it.

"So on one hand we told the environmentalists that if you care about the environment you have to learn to love hydrofracking. And on the other hand, we also told the industry that you have to get your act together."

The question, of course, is whether that can happen.

The vast power of the Marcellus and other shale plays has already been unleashed. So have the activists. And they now seem so intractably at odds that it's hard to believe that any accommodation is possible. But the costs of going on the way we are—those costs are staggering. And they won't be borne only by those of us who are struggling with the issue now.

Liam stumbled out of the Blazer and I followed him as he scrambled up the rocky incline to get a closer look at the furiously managed chaos at the crest of the pyramid. I found myself thinking about the field that used to be there, and about a sunny Sunday afternoon more than forty years ago. And about whether there were any lessons that I could draw about the Marcellus from what had happened on that field all those years before.

Not long after we bought the place, we bought our first horse, a beautiful palomino that my mother, enraptured as she was with all things western, and perhaps in part because she harbored a secret passion for Robert Redford, gave the cloyingly cute name of Sundance.

He was an impressive beast, but nasty. It was understandable. Though he was a gelding, he had been gelded far too late in life for it to have had any calming influence on him; instead, it seemed to have made him bitter and irascible and particularly hostile to the two-legged species that had subjected him to the indignity of it in the first place.

Unlike the other horses we had later, however, Sundance didn't express his resentment the old-fashioned way, by trying to buck you off. No, he was far more subtle. What he would do was run away with you, heading full tilt to the nearest apple tree, and no matter how hard you pulled back on the reins he wouldn't lose a step until he reached his intended destination, where he would then scrape you off his back as violently as possible.

I was all of about twelve years old at the time—a scrawny, transplanted suburban kid who didn't have the first foggiest notion of how to handle a mean-spirited beast like that, and it wouldn't have mattered even if I had. Far better riders than I had tried to best Sundance, and just like me, they ended up in a bloody heap at the bottom of a tree trunk.

But my father, who never sat a horse in his life, was insistent. He demanded that I keep trying to beat Sundance into submission and every Sunday afternoon, right after we got back from church, he'd order me into my room to change my clothes, swapping out my Sunday best for the ragged, bloodstained shirt and jeans I had worn the last time I had tried to ride Sundance. He would then march me out to the field behind the barn where we'd gather up Sundance, saddle him, and brace ourselves for the worst.

It always ended badly. For me. And I came to dread Sunday afternoons like a court date. Beginning late on Saturday night, my legs would start to get wobbly and my stomach would start to churn with a cold, burning terror, but there was no getting around it. My father wanted me on that horse, and I was going to get on that horse. Sundance wanted me off, and I was going to be knocked off, and wind up on my ass, scraped and bleeding, trying as hard as a twelve-year-old could not to cry as I painfully got to my feet and limped back to the house.

This went on for about a year, I suppose, before my father, out of either pity or disgust, finally relented and agreed to sell Sundance. He called up our trader, a guy named Buddy Baldwin, and Buddy agreed to show up the following Sunday with his truck and his wife and partner, a woman who had been graced by fate with the best name imaginable for someone in her line of work: Winnie.

Maybe it was just a boy's budding machismo, maybe it was just rank stupidity, but when Buddy and Winnie pulled their truck into that field and began preparing to collect Sundance, I asked if I could try just one more time to ride him. Winnie gave Buddy a long, meaningful look, Buddy looked at my father, and my father looked at me and nodded.

I had no reason to think that this time would be any different, and as we saddled him up, that old familiar terror started to churn again in my gut. I hopped on his back and took the reins from Buddy. Sundance pranced a bit and then did it again, breaking into a dead run, heading straight for the apple tree as I choked back sobs and yanked as hard as I could on the reins, begging him to stop.

I'll never really know where it came from, but just as Sundance and I rounded the barn, something inside me snapped, and I let out a bloodcurdling yell from somewhere deep inside me that I had never heard from myself before or since. Almost without realizing it, I dug my bony teenage heels into that beast's flanks, and rather than pull back on the reins, I let them go slack and began whipping him as hard and as fast as I could, shouting the whole time, and when he tried to pull up at the apple tree, I wouldn't let him. I just kept him running. We must have covered half a mile at a full-out breakneck gallop, off our property, down across the next farm and the one after that, by the time I turned him and headed back. When we reached the field where Buddy and Winnie and my father were waiting, there was nothing left of Sundance but hooves and foam. At the bottom of the hill, I tugged gently on the reins and he obediently came to a halt. I prodded him forward, pulled back again, and again he stopped.

I had beaten him, just as my father had wanted. Not by resisting his instinct to run, but by using that very instinct against him. And then I rode directly to my father, hopped off the horse, handed the reins to Buddy, and said over my shoulder to my father with all the bravado a thirteen-year-old could muster, "There. Now you can sell the son of a bitch."

Could it be, I wondered, as I stood there in what used to be that field, watching my own son process his conflicting emotions over the scene that was playing out before him, that the two most life-changing lessons that this place ever taught me could be applied somehow to the challenge that it, and all of us, are facing right now?

Is it possible that the two sides in this bitterly polarized debate, people who in some ways are separated as much by their similarities as by their differences, could find a way of looking past those fractures the way my father and I had? Could we figure out a way to use the raw power that had been unleashed here, not just the energy in the gas but the crude instincts of the industry, and the driven activism of its opponents, to channel a response to the dual dangers of climate change and fracking and find a way forward?

Standing there on that hill that day with Liam, I realized had already made a decision. I was going to go out and see whether I could find an answer to that question.

The aftermath of Hurricane Sandy on Staten Island, NY.
Photo by Karen Phillips.

{ TWO }

Comfortable in Our Ignorance

IT WAS A BITTERLY COLD NOVEMBER EVENING IN one of those off-the-beaten-track towns with cobblestone streets. It was the sort of place where the main street is lined with gingerbread houses and twenty-first-century coffee shops in nineteenth-century brick storefronts where you feel like you could almost warm yourself in the glow of their windows.

I had been invited to speak about my own moral ambivalence about the politically charged subject of fracking and climate change at the local bookstore, a dusty, out-of-the-way place with groaning wide-plank floors worn smooth by generations of voracious readers, back in the days when there still was such a thing, the kind of place where even the light from the street seemed to stop to browse as it passed the sagging bookshelves on its way to the back of the store. I jumped at the opportunity.

I'd given a version of that talk a few times before, and it was always controversial. There were always a few in the audience who let me know in no uncertain terms that they believed the whole idea of global warming was a hoax perpetrated by goatskin drum-beating extremists bent on destroying the free enterprise system. They saw it as a thinly veiled plot to bring America's workers under the sway of a rapacious government that was only barely containing its socialistic lust for other people's property.

And there were usually a couple of people who saw any concession to the idea that we might have to feel our way slowly out of our carbon addiction as an utter capitulation to a creeping corporate conspiracy hatched by the Koch brothers, Halliburton, Monsanto, and the Scaife Foundations to foul our water with petrochemicals and taint our food with GMOs in order to turn us into zombie slaves willing to watch mutely as they pollute our environment, leave us destitute, and come to my house to snatch my eight-year-old son and work him to death in the twenty-first-century version of the same coal breaker my grandfather escaped from a century ago.

I'm usually pretty light on my feet when it comes to these talks. I can generally navigate between these two poles. And I usually am able to find enough shared interests on both sides to show that at least a partial consensus is possible, even if only remotely. I had no reason to doubt that I'd be able to do it there, too.

After all, the folks in that town clearly understood the stakes, I was sure. Just eight weeks earlier, that bookstore and that little town had been in the crosshairs of Hurricane Irene, a major, destructive natural phenomenon, a five-hundred-year storm, the second such storm to rake the region in a generation.

And Irene was only a pale shadow of the megastorm Sandy that would slam into the East Coast a year later as the second deadliest and costliest storm to hit the country in modern times.

Irene's bull's-eye missed my hosts and their town, but their neighbors upriver were not so lucky. What started out as Hurricane Irene, and became the costliest Category 1 storm in U.S. history,[1] caused more than $15.8 billion in damage, almost all of it in a narrow path that stretched from the Susquehanna River basin in central Pennsylvania to the eastern banks of the Hudson in upstate New York and into western Vermont.

To be sure, there was disagreement in the scientific community about how much the changing climate, rising ocean temperatures, and rising sea levels added to the fury of the storm. But among the scientists and analysts I had spoken to there was little doubt that if Irene was not a smoking gun that demonstrated clearly that our addiction to fossil fuels had already set us well on the road to disaster, it was at the very least a harbinger of things to come, a warning of what we might face if we failed to reduce the amount of carbon that we were pumping into the atmosphere at a breakneck pace.

Certainly, I figured, a close call like that would make this precious little river town fertile ground for a discussion about the wisdom of compromise.

At least that was what I thought until about midway through my talk, when a member of the audience stood up, and in graphic detail related an incident so horrible, so extreme, that it proved, at least in that person's mind, that those on the other side of the debate were so fundamentally evil that no compromise with them was possible, and that no sane person would even consider it.

The incident, of course, had never happened. If it had, it would have been on the front page of every newspaper in the world; it would have been on an endless crawl across the bottom of every cable news network screen; it would have been an earthshaking international event on a par with the Deepwater Horizon disaster in the Gulf of Mexico or the discovery of Pol Pot's killing fields in Cambodia.

But I could look into this person's eyes and see that the person was not lying to me. The story was related with the same fervent, passionate faith with which a fundamentalist Christian might recount the story of Noah's flood, or an avowed Marxist might expound on the glories of a proletarian dictatorship. As far as this person was concerned, every word was absolute gospel.

I'm not, by nature, a particularly confrontational person, and so rather than embarrass the person directly, I opted to ask a few probing questions. I made it to the third question, and then I saw a look of panic start to creep across the person's face. I actually felt a twinge of sympathy as the person slowly came to fear that the story didn't hold up under scrutiny, that the whole complex structure of the person's worldview, at least as it related to this issue, was built on air.

At that moment, the person stood up, marched to the exit, spun around, glared at me, and sputtered, before storming out, "I'm not going to argue with you. I am comfortable in my ignorance!"

My first reaction? To borrow a phrase from my Australian wife—I was gobsmacked.

My second reaction was to vow that one day soon I would launch a petition drive to have "E pluribus unum" removed from our money and replaced with "I am comfortable in my ignorance." Because that is who we have become.

You've no doubt noticed that—clumsily at times—I have not told you which side of the issue this person was on. Nor have I told you precisely where this occurred. I haven't even told you the person's gender. There's a reason for that. Polls have indicated that on the issue of climate change, as on so many others, gender can sometimes be a predictor of position. So can geography. I haven't even detailed the horrific, apocalyptic event the person referred to, and with good reason. These days, even our most fevered nightmares are partisan. The fact of the matter is that this person's actual position, forged in fear and fabricated into a stubborn dogma, is irrelevant. It could have been on either side.

What that person personified is precisely what political writers and sociologists have filled volumes detailing in recent years, how long-existing cultural fractures are being exploited by the most extreme voices on the right and left, how the cultural fissures that they force open have spiderwebbed throughout society, rendering us more divided, perhaps, than we've been at any time since the Civil War. The divisions have paralyzed or at least marginalized our institutions—government, academia, even industry—at the precise moment when we most need those institutions to help us collect the data and devise the strategies to deal with the maddeningly complex challenges we face. Those fractures are rough and jagged, and they sometimes seem very, very deep.

If you follow the major issues of the day—climate, to be sure, but all the rest of them as well—in the mainstream press or on social media, it's easy to believe that there is one great yawning chasm between the people on one side of the political divide and those on the other, and you could even be forgiven for lining up on one side of the divide or the other yourself. Most people, at least according to polls, do just that. But the truth is, the fractures that snake through our cultural and political landscape are not that simple. They are complex, compound fractures, and people on either side of the great divide often find themselves straddling a number of those fractures.

Nowhere is that more obvious than on the issue of climate. Partisans on both sides of the issue often claim that there are only two sides to the debate—as the most ardent climate change activists like to put it, you're either pro-science or pro–fossil fuel, and there is no room for middle ground, or as the anti-man-made-climate-change ideologues like to cast it, either you're a dupe for what Oklahoma senator Jim Inhofe called "the greatest hoax" ever perpetrated on the American people, or you're a patriot.

But, of course, it's not as simple as being pro-science or anti-science, as much as partisans on either side might wish it were.

Science knows what it knows. Sometimes it has an idea of what it doesn't know. Sometimes it has no idea what it is that it doesn't know. It's not a parlor game. It's not a fortune-teller on a boardwalk. It's a process of understanding that is never-ending; it is constantly evolving. If we as a species waited for the theory of evolution to be unassailably proved before we actually evolved, we'd still be swinging from the trees trying to avoid becoming a leopard's midafternoon snack.

And right now, the vast majority of scientists who study such things are warning, with a surprisingly high degree of certainty and an even more remarkable degree of unanimity, that the leopard is crouched and twitching.

Ten of the fifteen warmest years on record occurred in the past two decades, while not once did the monthlong average temperature fall below the twentieth-century average. That has prompted more than one commentator to note that by 2013, no one under the age of twenty-eight had ever lived through a month of below-average temperatures.

On average, the United States has seen temperatures rise by 1.5 degrees Fahrenheit since the end of the nineteenth century, with 80 percent of that increase occurring since 1980, according to the draft of the Third National Climate Assessment report released in 2013. One study, published in the journal *Science* in 2013, concluded that temperatures, after a cool snap that lasted a few thousand years, were now rising faster than at any time since the end of the last ice age.[2] Scientists estimate that over the next several decades, those temperatures will keep rising, perhaps adding another 2 degrees to the averages in the United States, perhaps as much as 4 degrees.

And while that might not seem like much, they warn, those rising temperatures are likely to contribute to even wilder extremes—making events like the years-long drought that has been driving ranchers from their lands in Texas and elsewhere the new normal. It's sometimes difficult to tease the signal out of the noise, but there is ample evidence that the changing climate is already taking a toll. The ice caps are threatened. Parched conditions in the Mountain West are exposing timberland to increased pressure from pests like the pine beetle, and that, coupled with a host of other issues, risks turning vast forests into tinderboxes. And in the Midwest, America's breadbasket, wild extremes of weather, driven in part, most scientists agree, by the increasing temperatures, are threatening crops with floods one year followed by drought the next.

By the end of the century, scientists warn, many of America's coastal communities—the very places where the bulk of our national wealth is located—will be under ever-increasing assault from rising tides and spectacular storms like Sandy, the 2012 "superstorm" that ravaged the East Coast.

Peering further into the future, they warn that some of our cities—like Miami, for example—could become watery ruins unless the process is reversed. And even now, rising ocean temperatures are affecting the chemistry and biology of the sea with potentially devastating consequences, not just for the ecosystems of the oceans but for the economic systems that depend on them.

None of this, of course, is happening in a vacuum. Development, coupled with our rapacious consumption of everything from land to the resources beneath its surface, has led us to build in harm's way, while policymakers who are either unwilling or unable to grapple with the consequences of a changing climate have so far taken only halting steps, and those reluctantly, to acknowledge the challenge we face.

And to be fair, there is disagreement among some scientists about how much we have contributed to the changes in the climate. Natural variability, to be sure, has played a role in the weather extremes we've experienced in recent years. The drought in Texas, for example, or a killer heat wave in Europe would probably have occurred regardless of whether we continued to pump planet-warming greenhouse gases into the atmosphere, though there are few scientists who would argue that these would have been as long or as deep or as destructive as they have been.

But when it comes to the big picture, the science is becoming clearer, and despite the protests of some contrarians and the reluctance of many non-scientists to accept it, a consensus has formed that the climate is changing, rapidly, and that there are things we can do to reduce at least the portion of the problem that is of our own making.

We can't just snap our fingers and make the threat vanish.

Science, after all, is not magic. It is not the job of science to give us the answers we expect. Its job is to sharpen our questions, and hopefully to lead us to ask the right ones. It's a tool, a tool that allows us to use the best information available to make the best possible decision at the moment. Sometimes we have the luxury of allowing the science to ripen. Sometimes we do not.

Unfortunately, science is also the only tool we have to read the clock. At the moment, the preponderance of scientific evidence suggests that we have a limited amount of time to deal with the most pressing crisis we face. There is evidence that the climate, saddled with the growing burden of waste produced by our efforts to provide fuel for an increasing and increasingly affluent and demanding world population, is becoming increasingly volatile. That raises a legitimate concern that in the very near future, the instability of the climate could reach a point where we no longer have an ability to manage it.

Time and again, the political will to recognize and address the challenges that a changing climate poses, not just as an environmental issue but as an economic and even national security imperative, has fallen into the ever-deepening and widening crevasse between the two extremes.

On one side, there's the "stand your ground" crowd—those who advocate doing nothing in the face of this challenge—which defends inaction by declaring that there are still holes in the science and we should wait until all the answers are apparent before doing anything.

Those who are persuaded that the risk is real enough, and likely urgent enough, to require immediate steps see that as cynicism, using science's built-in uncertainties as a cudgel. And they accuse their opponents of being anti-science in large part because of it.

But they're not blameless either. While science has gone a long way toward alerting us to a threat, neither it—nor the free market, for that matter—has provided us with many options to avert or even manage that threat.

While we have made remarkable progress in energy conservation and deploying renewable energy, we're still a long way from the scale that would help us meet the equally critical challenge of providing enough clean energy to allow all of us—including the 1.5 billion people in the world who have no electricity at all or the roughly 2 billion more people the world is expected to hold fifty years from now—to live in dignity and a modicum of comfort. Perhaps, with appropriate support sooner rather than later, we will be able to develop those resources.

But for the moment we have to live with what we've got, using the least of evils to create the greatest of goods. Unfortunately, for many of us who are persuaded that the risk is real enough, that's a deal with the devil. It's a nonstarter.

Science is not supposed to be politics. Science—good science, used correctly—can identify risks. It can then help identify a range of possible responses to those risks. In short, science informs judgment and judgment defines action. It's a scalpel, not a battle-ax, and it should offer a menu of options, not a manifesto. But it is being used as a manifesto, hijacked by partisans on both sides to form the basis of two competing narratives, the grimmest of Grimm fairy tales, bedtime stories of ravenous oil-sodden monsters lying in wait to despoil the earth, or radical Maoists scheming to drown the American Dream in a gray pool of socialist absolutism.

These battling dogmas seem to offer no room for compromise. And the result is that we've fought each other to a virtual standstill. Claiming the moral high ground and building barricades out of selective bits of science may feel really good, and boasting about our own intellectual and cultural integrity compared with those extremists on the other side might win you the prized spot closest to the police barricade at a Tea Party rally or a White House protest over the Keystone XL Pipeline, but it's not going to make it any easier to develop a reliable way to reduce this nation's crippling dependency on dirty energy. Nor is it going to help build the kind of consensus needed to pull any kind of carbon reduction policy out of the political tar pits that such legislative efforts have sunk into three times in the past decade alone.

The problem, of course, is that the issue seems to fit so neatly into preexisting hyperbolic orthodoxies of the two predominant strains of American culture. As the authors Mathew Barrett Gross and Mel Gilles put it in a 2012 piece in the *Atlantic*,[3] the very factors that scientists believe are driving global climate change—an unalloyed capitalism reliant on the continued and accelerating use of fossil fuels—parallel the economic dreams of many conservatives. Meanwhile, the proposed responses that tend to bubble up in the media from progressives and many in the environmental movement tend to be every bit as contentious, calling for dramatically curtailing the power and authority of industry, for example, or moving toward a more localized trade system. It's little wonder, then, that the issue of global warming has "become just another front in the culture wars," as Gross and Gilles wrote.

Of course, it may not be quite as black and white as all that. While partisans on both sides continue to try to round up as many Americans as possible and dump them by the truckload into one camp or the other, the truth is that most of us fall somewhere between those two extremes.

We don't fit neatly into either of those two Americas. We fit more neatly into what the Yale Project on Climate Change Communication and George Mason University have called, in an ongoing study first released in 2009, *Global Warming's Six Americas*.[4] On the basis of polls conducted periodically over the last several years, the researchers concluded that most Americans can be sorted into one of six groups when it comes to climate change and the role of humans in it. The most concerned group, dubbed the "Alarmed," is, according to the 2012 poll, at 13 percent only slightly larger than those who fall into the category of least concerned, the "Dismissive," those who reject the whole idea and account for 10 percent. Most of the voices we hear in the public debate tend to fit into one of those two camps. The most strident advocates—the ones who believe that we face an impending cataclysm, who are actively making changes in their own lives and who demand that the rest of us immediately scale back our production of carbon and other greenhouse gases even if that means a radical restructuring of the global economy; or conversely, those comparatively few voices who argue that it's all nonsense and we should do nothing but expand our consumption of fossil fuels—represent less than a quarter of the American people.

In fact, the majority of Americans, 51 percent, according to the 2012 poll, are either "Concerned," which means they take the risk seriously and support some degree of government action to stem the tide of climate change but have not yet become deeply involved in the issue, or "Cautious," which means that they're not sure whether climate change is real—about a third believe it's real, a third don't believe it's real, and a third don't know—but tend to think that even if it is real, either it won't have serious consequences until well into the future or that we're already doing enough to address it. Thirteen percent are "Doubtful," which means that they believe that even if the climate is changing, it's not our fault.

Another 6 percent fall into the "Disengaged" group, which means they are utterly uninvolved.

And yet, despite the fact that most Americans are more nuanced in their positions than those in the "Alarmed" or the "Dismissive" categories, the public debate tends to be dominated by the most extreme voices.

There are a number of reasons for that, said Michael Mann, the climate scientist who has been the lightning rod for some of the most heated debate in recent years. The increasingly acrimonious tenor of all debate in this country—a phenomenon that has been almost institutionalized by a concerted effort to redraw electoral maps to create ideologically pure political ghettos on both the left and the right—has had an impact, giving extreme constituencies a greater voice in the public debate than their numbers seem to warrant. That ideological fracturing has affected the debate on all socially divisive issues, but particularly on climate, he argued.

In essence, the complex public debate has been reduced to two camps: those who deny that the climate is changing or that we're responsible for at least a significant part of that and those who believe that while the threats are mounting, we will have time to use the tools at our disposal—government, industry, science, academia—to develop a strategy to slow the march toward the 3- or 4-degree Celsius increase in global temperatures that most scientists believe would be disastrous.

That group of believers includes those who are convinced that it's not good enough to avoid reaching what scientists believe is the bright red line that marks concentrations of carbon at 450 parts per million in the earth's atmosphere. These are people who demand a return to 350 parts per million (we crossed the 400 threshold for a full day for the first time in millennia in May 2013), even if that requires a revolutionary overhaul of the planet's already struggling economic system, a system that is barely serving billions of the 7 billion people on earth, and will sputter even more as that total edges closer to 9 billion in the coming decades.

The increased balkanization of the media, with conservatives drifting toward conservative media outlets and liberals toward the liberal outlets, has also contributed to the problem, as has the media's propensity to, in the interest of balance, present two sides, and usually only two sides, of a multifaceted issue and then call it a day. The media, Mann said, have been largely "unwitting" accomplices in the muddying of the debate over climate. He used the word "pawns."

Though there are an infinite number of numbers between 0 and 1, the press in the United States, especially, tends to be a binary business, he argued, designed to approximate balance by getting two points of view on any subject even when those two points of view are not symmetrical. As a business, it is driven by market forces to book the voices that are the most extreme, and therefore the most entertaining.

To be sure, says Candis Callison, an assistant professor of journalism at the University of British Columbia who has built a career studying the intersections between scientists and the media, in recent years there has been a marked decline in science writing and reporting at every level from the national press all the way down to local newspapers. Chalk it up to a mad rush to cut newsroom costs, or a general decline in the standards of what passes for news, but the end result is that there's a lot less science expertise in newsrooms around the country. And while there has been, at the same time, an explosion in the number of specialized, boutique publications and websites that focus on the interests of climate geeks, those outlets tend to preach to the choir.

What's missing, she says, is the old-fashioned voice-in-the-public-square role that newspapers and local radio and television outlets used to play.

"You've seen science sections shut down," she told me. "You've seen them pretty much disappear in local and regional newspapers. . . . In general, there's less opportunity for journalists to pitch those kinds of stories."

And when the stories make it into the newspapers or the local six o'clock news, the hidebound traditions of the craft—the need to tell a story in a top-down "inverted pyramid" form familiar to the readers, with competing experts offering conflicting interpretations—often does as much to obscure the facts as to illuminate them.

One result is that the nuances that should define the debate are lost. Another is that when the public hears about nothing but the extreme positions, the potential to reach common ground is eroded.

"I'm a big fan of Bill McKibben," Mann said, referring to the writer and activist who was among the first to write about the whole issue of climate change for a popular audience and who has in recent years made it his mission to champion the cause of clawing our way back to a CO_2 concentration of 350 parts per million. McKibben even named his organization 350.org. "I accept some of the scientific reasoning behind his position that the only truly safe option is to get CO_2 back to where it was when you and I were growing up," Mann said. "Nonetheless, that is a far more steep uphill battle than simply stabilizing CO_2 concentrations.

"I think it's okay to have that voice out there because it's a valid view. But if it becomes too dominant in the discussion, there's a danger that it makes it more difficult to meet in the middle with respect to policy positions to get us on the right track. . . . If you tell people we have to get to 350 and they already see how difficult it would be to even stabilize at 450 . . . when people see what an uphill struggle that is, and you tell them, 'You know what? That's not good enough,' it's an excuse for denial, an excuse for throwing up your hands in defeat and saying, 'Okay, there's no way I can do it anyway, so I'm going to continue driving my Hummer and the profligate lifestyle I'm living because we're going to hell in a handbasket no matter what.' I think there is that danger."

The sheer, sometimes terrifying magnitude of the challenge, he argued, has been fertile ground for exploitation, particularly by those interested—usually, he maintains, for very personal financial reasons—in blocking action on a federal, state, or sometimes even local level to reduce greenhouse gas emissions or adapt to the changing environment. Those forces, the side that's trying to stymie action and the side that demands immediate action, have turned the issue into what he calls "an ideological litmus test."

The problem with that response, in Mann's view, is that it strengthens the hand of the do-nothing side. "I think it behooves whoever is on the side of inaction to polarize the public on the issue," Mann said, "because you don't need to win the argument to prevent public policy, you just need to polarize the public."

He cites the role the movie *An Inconvenient Truth* played in the fracturing of public opinion as a case in point. "You hear people say that movie polarized the public, and that's not quite a correct reading of what happened," Mann said. "That movie was an extremely convenient vehicle for an interest that was looking for a way to polarize the public.

"Al Gore played into that. He was a partisan political figure. He became the face of the issue and they used that to divide the public," he said. "Gore was an unwitting participant in that phenomenon. But from that point forward, the public did become more polarized. And now it is a litmus test, certainly within the conservative movement. There are a set of positions you're expected to take and one of them is to deny that climate change even exists. It's written into the Republican platform."

It would be enough of a challenge if that harsh partisanship affected only harsh partisans, but that's not the case. At least that's what Yale researcher Dan Kahan has concluded. Kahan, a professor of law and psychology, says that even though most people know comparatively little about the science behind climate change—after all, they have lives, why should they be expected to immerse themselves in the mind-numbingly complicated data?—they do tend, simply by virtue of living in America, to know more or less where they stand on the issue. They stand with the guys they trust on other issues.

"When you talk about climate change, for most people it's not something they talk about very often. But it is something about which, at this point, they're going to know what their position is. Because the person that they are and the people that they hang out with kind of cohere with a certain position," Kahan said. "If you talk to them about it, they'll have the position. But I don't think they'll . . . relish the idea of having a conversation about it. It's kind of . . . a tense issue. . . . They have these positions, but it's not like they studied it. And they don't want to get into a debate about it. They believe what they believe."

It is, he argued, an evolutionary adaptation that has on comparatively rare occasions led us to plunge, lemming-like, over a cliff right behind some demagogue, but by and large it has served us pretty well, this ability to sniff out who among us both understands complex issues and can put them in their proper context for the particular tribe we belong to.

"The most remarkable thing about human rationality is that we're able to kind of pool information that we have and build on it," he said. "I mean, how else could people participate in all this massive knowledge that existed if we didn't have a way—short of understanding it ourselves or producing it ourselves—to know who knows what about what?

"In fact we wouldn't have much information to begin with if we didn't have that," he said.

The problem is, by gravitating to the sources we're already comfortable with, Kahan said, "we end up, I think, with an exaggerated sense of the polarizing character of this because we're ignoring the denominator. We're not griping, in this country at least, about pasteurized milk. We're not arguing about whether the power lines are causing people's brains to explode. That's not because we know more biology or electromagnetism or whatever the heck it is that you'd have to know. . . . It's the networks we're using to get the memo. We're using these networks of who knows what about what and they're leading us to the same place."

Thus, certain issues, for whatever reason, become cultural markers. Climate is one of those issues.

In a way, Kahan said, it's rare for an issue to reach the kind of critical cultural mass that climate has. He used the word "pathological" to describe it, the idea that people "are going to have this tremendous kind of investment in holding on to that belief. And that's going to drive them apart."

We let our proxies do the talking for us. Increasingly, they're becoming more polarized, and to rise above the din, their voices are becoming increasingly apocalyptic.

{ THREE }

Kindergarten in a Fallout Shelter

I WENT TO KINDERGARTEN IN A FALLOUT SHELTER.

It wasn't a big deal. A lot of people did back in 1963, the year I started school. After all, it was only ten months after the Cuban Missile Crisis, a time when the world was divided roughly into two hostile camps, both sides bristling with enough nuclear weapons to vaporize most of the people on the planet several times over, and all of us had just gotten an object lesson in how perilously close we had come to tumbling into the abyss of a nuclear holocaust. And so, when my parents enthusiastically escorted me to my very first day at St. Paul's Grammar School and walked me down the steps into the basement where Mrs. DelMastro's classroom was, we were greeted by her warm smile and by a room all decked out with cardboard cutouts of cheerfully cartoonish cherubim, giddy anthropomorphic lambs, the big block letters of the alphabet, and—on a bright yellow metal placard placed directly above the door—the international symbol for radiation.

We didn't talk about it a lot, but for those of us who grew up in the Cold War era it was always in the background, as much a part of our elementary school experience as the smells of crayons, pencil shavings, and souring chocolate milk. It was in the air, this low-level, throbbing sense that at any given moment, some barely noticed geopolitical butterfly might be flapping its wings in some remote corner of the globe and could set in motion a complex cas-

After Sandy. A woman poses with one of her few remaining possessions, her wedding dress. Photo by Karen Phillips.

cade of events that might escalate into an all-out war between superpowers, a final, apocalyptic showdown that would kill millions instantly and consign the rest to a brutal, zombie-like existence in a dying radioactive hellscape under the roiling gray skies of a nuclear winter.

Don't get me wrong. It wasn't uppermost on our minds. We still did kid things. Kickball. Freeze tag in the macadam parking lot that doubled as our playground. It's just that sometimes when you got tagged, you'd look down at your shadow on the ground and maybe judge your pose a little bit by how you imagined it would look etched there forever by a thermonuclear flash.

Over the next twenty years or so we got used to the idea that we were always one bureaucratic miscommunication away from the End of Days. We went on with our lives, of course, went to high school, dressed up for the prom, got our driver's licenses, went to college, got married, had kids, and got buried under the mountains of daily obligations that come with adulthood, And in the meantime, the nuclear apocalypse failed to materialize.

There were, it turns out, enough good people on both sides of the bipolar globe—motivated by altruism or by self-interest or, more likely, by their managers a few government pay grades up the food chain—working to ratchet down the tensions and to maybe try to reduce the number of armed warheads we had pointed at each other, to avert disaster. Sure, they had to contend with domestic political interests that worked feverishly to exploit the preexisting fractures in society, those on the one side who claimed that the only proper response to deadly nukes was even more deadly nukes, and those on the other who, in an abundance of faith in the innate goodness of human beings, insisted that the only appropriate response was immediate disarmament, unilaterally if necessary. It wasn't easy, and it would be naive to believe that the real danger has been eliminated. But it has been contained and reduced, and by the time I took my eldest daughter to her first day of kindergarten in 1996, those bright yellow international radiation signs next to the smiling lambs were as much a part of Daddy's distant past as record players and heavy plastic telephones hanging on the kitchen wall.

There is, however, precious little evidence that the tendency toward apocalyptic thinking that was so much a part of the ethos when I was a child has abated, and nowhere is that more evident than on the issue of climate change.

As with the nuclear danger in the 1960s, the potential risks of global climate change are staggering. The worst-case scenarios, assuming that we continue to churn out carbon and other greenhouse gases with abandon, are truly horrifying: seas that could rise by yards rather than feet; unimaginably vicious storms that could scour densely populated coasts clean of people; noxious gas long trapped in the frozen north released, escalating the cycle; the ice caps gone; savage, almost permanent droughts turning vast swaths of the planet into desert, resulting in global instability, hunger, disease, and war. And that's just the tip of the melting iceberg.

To be sure, the majority of scientists tell us, we are already seeing the early warning signs: rising sea levels, record low levels of ice in the Arctic, perhaps dangerously warming deep water in the Antarctic, droughts, wildfires. But within the range of possibilities and probabilities, there are other potential scenarios that, while still challenging, need not necessarily be the end of the world as we know it. These are dangers, these scenarios tell us, that are real, but we may have the wherewithal to manage, and perhaps someday even reverse the damage, if we have the will.

The problem, of course, in a debate that tends to be dominated, at least in the press, by the two ends of the spectrum—the Yale researchers' "Alarmed" on the one end of the spectrum and "Dismissive" on the other—is that more-modulated estimates seldom get much airtime.

One of the things that I've found while traveling around the country talking to people about climate change is that if you scratch people who adhere to either extreme, you'll often find an apocalyptic vision of the future. Either it's a left-wing vision of a dystopian society of haves versus have-nots, where a suffering remnant is struggling to stay alive on a planet denuded of all of its resources by greedy corporatists, or it's a right-wing nightmare of slavish automatons staggering through a cartoonish socialistic world of deprivation and despair. The only common ground between these two visions seems to be that in both cases, the desolate streets of the future are prowled by a disproportionate number of sweaty, muscular men in leather, right out of *Mad Max Beyond Thunderdome*—which may tell you much more about the puritanical past of our apocalyptic iconography than about what the future may really hold.

It is perhaps no surprise that a nation run by people of my generation, people who were raised with an expectation of annihilation, however muted it might have been in our day-to-day affairs, is drawn to apocalyptic imagery, say Mathew Barrett Gross and Mel Gilles, authors of *The Last Myth: What the Rise of Apoc-*

alyptic Thinking Tells Us About America. Our popular media, they say, are rife with pseudo-documentaries about one pending apocalypse or another. Our cable networks, no longer satisfied with turning sharks into fodder to ramp up fear for a week every summer, have ratcheted up the terror and feed us freeze-dried end-timing on shows like the National Geographic Channel's *Doomsday Preppers,* while the History Channel added Armageddon Week to its lineup. Partisans may think that simply to pierce that noise and be heard, they have to be more end-timer than thou. As climate scientist and evangelical Christian Katharine Hayhoe told me recently, there seems be a tendency to begin all discussions about climate from a starting point of fear. To be sure, fear can be a powerful motivator, she said. But it can also cripple us rather than prodding us to action; it can leave us feeling helpless. It's the high-fructose corn syrup of emotions. "The problem is that fear does give us a quick, short-term motivation. It gives us a shot of adrenaline," she said, a shot that media, which she argues have geared themselves to pander to an audience that craves fear, are only too happy to package and sell back to that audience. But it's short-lived.

The maxim behind it all is, as Richard Nixon once put it, simply this: "People react to fear, not love," Mathew Gross told me. That, he argued, is particularly true in a country like the United States, where apocalyptic thinking has been woven into the fabric of the national character since its inception. This is a country, he argued, that was founded by people who fled the strife of Europe after that continent's religiously induced end-times spasms in the seventeenth century and came here, at least in part to usher in the other aspect of apocalyptic thinking, millennialism—the idea that after the spasms, there would be a thousand years of peace and prosperity. Of course, that end-times era didn't happen any more than the Russian-launched nuclear holocaust that I was raised to dread did. But every now and again in our history, a new vision of the same scenario crops up, burns brightly for a while, and then fades out.

That aspect of our national character is a singularly rich vein to mine if your object is to advance a particular political agenda. It's very effective—and it works on both sides—to appeal to the underlying cultural touchstone that basically suggests that you're part of an enlightened or redeemed minority, that those who oppose you are not only agents of the coming disaster but will eventually be cast into perdition (be it a theological vision of a thirsty hell or a secular version where the other side is just as thirsty, but was enlightened enough to dig water wells and power their pumps with windmills or solar panels).

As a strategy to prod people to action, though, fear seems to come up short, Gross said.

Part of the problem is that the very real dangers posed by the rapidly changing climate just get drowned out, he said. The real dangers become just another apocalyptic scenario in a media environment already bursting at the seams with such dire predictions, just more white noise. "The danger of the media conflating apocalyptic scenarios is that it leads us to believe that our existential threats come exclusively from events that are beyond our control and that await us in the distant future and that a moment of universal recognition of such threats will be obvious to everyone when they arrive," Gross wrote.

Of course the challenges presented by a changing climate don't arise suddenly, not like a meteor hurtling toward the planet Earth in some adventure movie. They happen by degrees—a little higher sea level, a slightly warmer atmosphere, a series of storms more severe this year than last, worsening measurably but incrementally over time, giving us time to adjust, if not our infrastructure or our consumption habits, at least our expectation of what "normal" is.

The dire warnings about the worst-case scenarios, even if they are well within the range of possible outcomes that we might expect, have thus far failed to motivate public opinion on the issue of global climate change, Gross argued, but instead, perversely, those warnings may actually be contributing to the sense of inertia. Those among the Alarmed who paint pictures of global devastation to come—flooded cities, abandoned wheat fields, scorched mountains, starving masses—may inadvertently be giving ammunition to the Dismissive, who can point to every tree that's still green, every street that's not flooded, and argue that it's all hype. As early as 1989, a year after James Hansen effectively fired the starting pistol that began the great climate debate, Patrick Michaels, a climatologist and climate skeptic, warned of the danger of what he called "apocalyptic environmentalism."

Even more perversely, there is some evidence to suggest that, at least among some true believers on either side of the cultural barricades, the more fraught the debate becomes with end-of-the-world imagery, the less likely they are to support taking specific steps to address it. A study based on polls taken in 2007 and published in the journal *Political Research Quarterly* in 2012 by researchers David Barker of the University of Pittsburgh and David Bearce of the University of Colorado found that 56 percent of Americans believed in the Second Coming of Jesus Christ and that the belief in the end-times made them 12 percent less likely to strongly support government action to stem greenhouse gases.[1]

The way Barker explained it to me, some portion of that 56 percent of Americans who believe literally in the end-times are also perfectly willing to accept the idea that the climate is changing and that we're hurtling toward a global catastrophe. But unlike a lot of the rest of us, this group actually seems to welcome the idea. "We were trying to figure out," said Barker, "why it is the case that of all of the people who were opposed [to strong government action to slow climate change] a lack of belief in the phenomenon didn't explain everything."

What they found, of course, was that many of the participants in their survey were true skeptics who simply did not believe in man-made climate change. "You might be opposed to a policy to raise emissions standards . . . because you just don't believe in global warming. . . . You don't trust the science, . . . you think it's bogus," Barker said.

But there was another group, he found, people who were willing to accept that the climate was changing, even willing to accept the idea that the changes posed a potentially catastrophic threat, but who "still weren't terribly interested in doing anything about it."

That was puzzling, Barker said, but as the researchers dug deeper into the data, they hit on a theory. As it turned out, "a significant part of that actually can be explained by folks who believe in the Second Coming of Jesus Christ and believe that that Second Coming might be relatively imminent and so, with the idea being that, hey, if the whole world is going to end relatively soon anyway, and it's essentially going to end in the way that we like—with Jesus coming back—then why in the heck should I be wanting to have to pay more for a car and pay more for gas and stuff like that to try to fight something that doesn't matter to combat that?"

Of course, he argues, there is no evidence to suggest that the extreme religious right has a monopoly on dogmatically apocalyptic thinking. While he hasn't yet conducted a formal study of the far left, his personal experience suggests that "when it comes to apocalyptic thinking . . . the left seems very, very, very . . . willing to readily embrace the idea that, 'Oh, my God, everything is going to explode in a few years because I've seen Al Gore's movie.'"

There may, however, be a somewhat less despairing aspect of Americans' apocalyptic thinking, as I learned in a Sunday-morning telephone chat with Boston University historian and millennialist expert Richard Landes, author of *Heaven on Earth: The Varieties of the Millennial Experience.* When I caught up with Landes by phone, he was doing research in Jerusalem, a city that has seen more than its fair share of millennialists during the past couple of millennia.

The way he sees it, the debate between the most extreme voices on either fringe, amplified as it is by media that tend to see things only in sharp contrast, fails to grasp a fundamental truth about the peculiarly American brand of apocalyptic thinking. Unlike the European and Levantine variants that spawned it, he argues—which tend to be fatalistic, particularly on the issue of climate—there is, in the American version, a deep undercurrent of hope.

"I would call it 'avertive apocalypticism,'" he told me. It's rooted in a notion of redemption and fueled by the other unique characteristic of Americans, the idea—or the myth, if you prefer—that there's nothing we can't do. It's a driving belief that the worst aspects of climate change can be averted, that "if you change your ways, you'll be spared the catastrophe."

It may be true that more Americans believe in angels than in the concept of anthropogenic global warming, but, Landes argued, the angels they believe in are, more often than not, what Lincoln called "the better angels of our nature."

"The real question is do we sit around and wait for the angels or are the angels helping us to take action?" There are, he argued, two ways of approaching a pending apocalypse. The first, and the one that gets the most attention in the press, is the one that depicts the event as something so large, so beyond our ability or willingness to respond to, that it prods us toward inaction. The passive scenario, he calls it. The other is the one we glorify in our movies and our literature, in the stories we tell ourselves about our past—the active scenario, the one in the movie where the meteor is racing toward Earth and we send up Bruce Willis and Ben Affleck to blow it up with one of those nukes we have left over from my days in kindergarten.

"Most Americans don't think along passive lines," he said.

The sense of hope I heard from Landes that morning, so utterly different from the doomsday declarations I hear all the time from my friends on both fringes of the climate debate, was in many respects precisely what I had heard from many of the same scientists who have been compiling the data that now show how serious and potentially world-altering the challenges we face are.

As we'll see later in this book, these are not people who are prone to underestimate the dangers. And yet, by and large, they seem to have great faith in our ability—through mitigation and adaptation—to survive.

The way they see it, our history is the history of a people who've been dangling their feet over the rim of a volcano and yet somehow have always managed to survive. In fact, more of us are doing better now than ever before. We're taking our toll on our world, to be sure, and without some profound changes, we could most certainly trigger the kind of apocalypse that most of us fear and some of us long for.

But we probably won't. Maybe that sense of optimism resonates with me because of who I am and when I became it. I'm a guy in his mid-fifties who began life in a world that seemed destined to vaporize itself in a nuclear cloud but for some inexplicable reason didn't. As one scientist put it to me, "It's fairly clear that we don't know how to kill ourselves off even if we wanted to. We are the ultimate weed."

We know what we can do, we know what we must do, and we know what we will do, however reluctantly. So the question really boils down to this: Can we alter the tenor of the debate and create a political environment open enough to discussions of the natural environment to take concrete steps before we're forced to? I don't know. But just as there were people working behind the scenes when I was a kid to make sure that we didn't blast ourselves off the planet with thermonuclear warheads, there are people working now to open up those lines of communication and to build consensus on the issue of climate.

{ FOUR }

Preaching to the Choir

IT'S NOT THE SORT OF THING I TALK ABOUT A lot. It is, for me, a singularly private matter, but for most of my adult life, I've been playing a game of hide-and-seek with God, and for the better part of that time, God has apparently been playing by Rodney Dangerfield's rules: "I used to play hide-and-seek. They wouldn't even look for me."

Friends of mine have been a lot more successful at the game.

I had an old buddy many years ago, a dogmatic, hard-core atheist who believed that he could sniff out traces of creeping religiosity everywhere he looked. I envied him. "I've spent most of my life looking for some trace of God and I've had to settle for hints," I told him. "But you? Man, you see the son of a bitch hiding behind every bush."

Others, guys I knew growing up at the farm, for example, who had been raised in a strict fundamentalist tradition, also had a gift for absolute certainty that I both envied and, if truth be told, pitied. I rarely said it out loud, but I always sort of felt sorry for people who believed that God punched a clock six thousand years ago, created the world in six days, and has been biding his time ever since, doing occasional parlor tricks like sending plagues of locusts, splitting seas, stuffing Jonah into a big fish, and every once in a while pulling off a resurrection just to keep his skills up. I always assumed they missed the majesty of what I always took to be an ongoing act of creation. But I envied their absolute certainty, even if, from time to time, I found the expression of it puzzling. I remember once, while putting in hay at the farm with my father and a neighbor, a guy who had only recently been washed in the blood of the lamb, so to speak, we had, perhaps injudiciously, allowed the conversation to drift off onto matters of faith, always a risky subject when you're a member of one of the few Roman Catholic families in a predominantly Protestant neighborhood. "You know," my neighbor said to my father, "you and your wife are such good people. It's a real shame that you're going to hell."

I suppose that like many Americans, I had come to believe that faith was one of those unbridgeable divides in our society, that the closer you were to your specific vision of God, the further you were from the practical realities of the world. I was lucky enough to have had that belief shaken during my decades of hide-and-seek. It was more than twenty-six years ago, and I was hiding inside an eruv around a strictly Orthodox Brooklyn neighborhood with a friend of mine, a sometimes inspired, sometimes borderline-insane rabbi. It being late on a Sabbath afternoon, he and I had wandered into a storefront synagogue frequented by a local Hasidic sect just in time for the Third Meal. The place was jammed, of course, with the women, their heads covered, hidden behind a *mechitza*, a flimsy wood-and-muslin divider, and the men, all decked out in their lush fur hats, their silk coats and stockings, rocking to and fro at long tables, singing an ancient melody in the most passionate, mournful, and yet somehow joyous voices. My rabbi friend, with his plain wool yarmulke and his pure wool business suit, seemed a little underdressed. Me, with my off-the-rack blended suit from Men's Wearhouse, my borrowed yarmulke, and my red ponytail stretching down to the middle of my back—well, I just looked wildly out of place, and I felt that even more acutely when I noticed that from across the room, a middle-aged Haredi Jew was staring at me. I tried to ignore him, but I could feel his eyes piercing me from underneath his *shtreimel*, as he sat there, twirling his *payos*, his sidelocks of hair. My sense of dread only increased when he stood up and strode purposefully across the room toward me.

"Oh, hell," I thought. "Here it comes. He's going to give me shit for mixing the fibers in my suit."

He stood just inches from me. And then, he opened his mouth. I was stunned. The guy sounded like every middle-aged burnout you've ever met, like Cheech Marin in a pair of silk knickers. "I really like your ponytail, man," he drawled.

"*What?*" I said.

"Yeah, dude. I used to wear my hair like that when I was hitchhiking through Europe."

I've found myself thinking back to those experiences a great deal lately in the context of climate change. There is a tendency—it's certainly one I have—to imagine that there is an impenetrable *mechitza* between the world of science and the world of faith, that facts are facts, faith is faith, and never the twain shall meet. That, of course, ignores the reality that the one often, and for many, informs the other. No one, for example, could accuse writer and activist Bill McKibben of being anti-science—the guy uses data the way the rest of us use dental floss. But McKibben often draws on both his faith and the scriptures to bolster his arguments, as he did in the spring of 2012 when he delivered a stem-winder of a sermon at New York's famous First Presbyterian Church, using the biblical story of Job to make his case that action on climate change was a moral, and, yes, a religious imperative.

When Penn State paleoclimatologist Richard Alley describes himself, he uses three words: "scientist," "father," and "churchgoer," and he proudly told me that if I was looking for him on most Sunday mornings I'd find him down the hill from his office at the local limestone Methodist church in downtown State College, belting out the old hymns. "We sing real well. It doesn't matter if you're a Catholic, it doesn't matter if you're a Jew, we Methodists almost certainly sing better than you."

But even among evangelicals and fundamentalists, people who tend to take a more literal view of scripture, there is a growing

Katharine Hayhoe. Photo by Mark Umstot.

chorus of voices arguing that rather than being an obstacle to accepting and responding to the risks posed by climate change, their faith should actually be an inspiration and an impetus to action.

Perhaps no one better exemplifies that growing chorus than Katharine Hayhoe, an atmospheric scientist at Texas Tech who was one of the reviewers for the Nobel Prize–winning IPCC report in 2007, contributed to the long-awaited 2013 National Climate Assessment, and is an author, a mother, and—oh yes, a devout evangelical Christian.

She has, to be sure, no illusions about the science. She can rattle off the statistics: how in the worst-case scenario the planet could be heading to a temperature increase of somewhere between 3.6 degrees and 5.3 degrees Celsius, and how that could be held—at least temporarily—to as little as 2 degrees Celsius with the proper mix of reductions in carbon emissions and other greenhouse gases and increased energy efficiency. She certainly appreciates the risks that even moderate temperature increases pose, not just in the future but now. But perhaps the greatest risk we face, she argues, is the risk that we will be paralyzed into inaction, either by fear or by a kind of enforced dogma. "I think a lot of our problems have a lot to do with fear," she said, "and a lack of hope."

That paralysis, she argues, is not just a risk to the stability of the planet or the health and well-being of its 7 billion people; it is also corrosive to the fundamental pillar of her faith. "Our society has an idea that there is fundamental conflict between science and faith. That's an idea that's been part of Western society since . . . Galileo . . . was persecuted by the Church," she said, adding wryly, "If Galileo had been a little more tactful there wouldn't have been a conflict."

For her part, Hayhoe believes there is no conflict between science and faith, and instead sees the two as complementary. She gets a bit poetic when she talks about it. "The universe operates according to physical laws that we can observe and deduce and understand. . . . In a sense, by figuring that out, I feel almost as if I'm looking at God's lab notes from when he made the planet."

But science cannot, she argues, stand on its own, and there is as great a danger when people try to substitute science for holy writ as there is when people try to substitute holy writ for science. "Sometimes," she said, "I feel like in the scientific community there's . . . a misconception . . . that science can answer all of our questions, that science can give us all of our solutions.

"Science tells us what is happening with climate change, and our Christian values—to love others as Christ loved us, to love our neighbors as ourselves and to care for creation—demand that Christians take action."

Her unique position as both a scientist and an evangelical Christian has, she contends, made her a kind of double evangelist, preaching on both subjects. That poses real challenges. Even among some of her scientific colleagues, her Christian background is cause for curiosity or worse. Among some in the evangelical movement—people who, she emphatically argues, have been coopted by political interests that don't have matters of faith foremost in their hearts—she sometimes encounters suspicion.

On that subject, she's reluctant to cast the first stone, but when she does, she aims it directly at Fox News, which she accuses of willfully distorting the issues surrounding climate, and she quickly follows it up with a second stone aimed at the forehead of the GOP, noting sweetly that she's found nothing in her Bible studies to suggest that Christians are obligated to support the Republican Party.

There is, she says, only one way to respond to the challenge of being doubted by both scientists and true believers. "You have to be genuine.

"In the scientific community you have to be a real scientist," she said.

"In your Christian community you have to be genuine there," she added.

That's particularly important in more conservative, fundamentalist circles, she said.

"If you are reaching into more conservative Christian communities, especially ones that have that perceived a divide between science and faith, then your scientific credentials are actually a source of suspicion. . . . You have to have the credentials for that community."

And so, when Hayhoe speaks to the Christian community, she makes sure they know where she stands.

"It could just be a couple of sentences, talking about our shared belief, how we believe that God made the world, how we believe that God gave it to people to care for, how God's creation is so incredible and awesome, and how I see examples of that in my work. And then we go on to talk about what we're seeing."

What she's found is that recently—and she concedes that it's anecdotal—there has been within the evangelical community a growing sense that something profound is changing, not just with the planet but also within the community itself. There is, she said, a greater hunger for hard information delivered by a person of faith, somebody who shares their values. And other preachers, people like Jim Wallis and Rick Warren, are also beginning to cast the debate about climate in Christian terms.

Until fairly recently, Hayhoe saw herself as a voice in the wilderness. But that, she says, may finally be changing.

{ FIVE }

Running from a Grizzly in Your Slippers

IT'S A LITTLE CORNER OF PARADISE, OR AT LEAST it seems that way to a guy who grew up hunting and fishing on the East Coast and who dreamed of someday living in Big Sky Country. It's a 25-acre slice of the great Northwest at the far western edge of Montana. There's a little house on the place, just big enough for Todd Tanner, his wife, and their eight-year-old son.

He doesn't really need much more, he told me. Not that he could afford much more even if he wanted it. He is, after all, a writer—his byline has frequently appeared in the bait-and-bullet journal *Sporting Classics*—and he supplements his meager income as a scribe by acting as a hunting and fishing guide. The fact of the matter is, Tanner's life takes place outside.

He always figured his son's life would, too. But lately, he's begun to worry. He's seen changes in the Northwest, changes that seem to be accelerating with each passing year. It doesn't seem all that long ago that Tanner would bundle up and head out hunting in temperatures that would dip to 30 below zero. "The last two winters, we haven't hit zero at the house." And all around him, "everywhere I look, I see dying trees."

Todd Tanner is certainly not a scientist. He's not a politician or a polemicist. You'd be hard-pressed talking to him to figure out what his politics are. "Depending on what the issue is, I have a re-

Todd Tanner.

sponse that may or may not surprise you," he told me, with a bit of pride. "I'm one of those guys who's all over the map."

He considers it a point of honor that he's not "easily categorized." But then he categorizes himself. "I'm really a hard-core sportsman. I'm fifty-two now, and I've been fishing for, I don't know, forty-eight years?" His earliest and most precious memories, he told me, were hunting with his father and his uncles and his grandfather on the neatly parceled farm fields around Pauling, New York. "I ate it up," he said, "and it's become my life."

It was, he said, more than thirty years ago when he decided to shake the dust of Pauling off his snake boots and head west. He found a bit of work guiding fly fishermen and hunters, who often as not are high-powered corporate types for whom the great outdoors is like another planet altogether. He augmented the meager earnings he got from that with the even more meager earnings from writing about the outdoors, mostly for sportsmen's magazines.

But it did permit Tanner to live where, and more importantly, *how* he wanted. And it also gave him a front-row seat from which he could see firsthand the changes that are already coming to the West.

Over time, it has also led him to become, like Katharine Hayhoe, a kind of voice in the wilderness. Lately, he said, he's compounded his financial insecurity by becoming one of a small but growing number of traditional outdoorsmen who have begun to recognize that the changes he sees outside his cabin are the first faint signs of something more serious, more potentially devastating, looming on the horizon. And it's turned him into an advocate for education, specifically among hunters and fishermen, about the risks of climate change.

"I think I wrote the first feature story on climate in a sporting magazine, maybe seven or eight years ago, saying this is a big issue and we need to pay attention to it," Tanner said. And in the years since, the signals have become even clearer.

Tanner has immersed himself in the science of climate change, and he is as quick as anyone to note that it's sometimes hard to separate the signal from the noise when it comes to the obvious manifestations of a changing climate. Referring to the stand of dead and dying pine trees outside his cabin, for example, he told me, "Can I say with one hundred percent certainty that every one of those dead trees is climate-related? No. Absolutely not."

But Tanner can also quote chapter and verse from a host of scientific studies concluding that the precious woodlands he calls home are already in jeopardy and could face disastrous consequences unless steps are taken to reduce the amount of carbon and other greenhouse gases that we emit. He can cite the reports indicating that the United States' 600 million acres of forest sucked up about 13 percent of all the carbon emitted in the country in 2010, from a low-end estimate of 7 percent of all the carbon emitted annually in the United States to a high of 25 percent, making it not only a playground for hunters, anglers, and campers, but a crucial carbon sink, according to the draft 2014 National Climate Assessment.

And he can take a certain joy in the fact that currently there is so much forested land; it is, after all, actually a kind of victory, albeit a small one. After 237 years of decline in forests resulting from industry and agriculture, the nation actually began to reverse the trend and gained a bit of woodland in the second half of the last century, much of it former farmland.

That's true across vast swaths of the country. But Tanner is also painfully aware that the same threats that have been linked to climate change elsewhere—drought, extreme weather events, the rise of certain types of insects (notably the pine beetle, which bores its way into standing pines in the West, kills them, and leaves them tinder-dry, providing more fuel for wildfires)—are threatening the forests, particularly in his corner of the Northwest.

According to the National Climate Assessment draft report, "Insect outbreaks and pathogens, invasive species, wildfires, and extreme events such as droughts, high winds, ice storms, hurricanes, and landslides induced by storms . . . are all disturbances that affect U.S. forests and their management. These disturbances are part of forest dynamics, are often interrelated, and can be amplified by underlying trends—for example, decades of rising average temperatures can increase damage to forests when a drought occurs. Factors affecting tree death, such as drought, higher temperatures, and/or pests and pathogens, are often interrelated, which means that isolating a single cause of mortality is rare. However, rates of tree mortality due to one or more of these factors have increased with higher temperatures in western forests and are well correlated with both rising temperatures and associated increases in evaporative water demand. These factors are consistent with recent large-scale die-off events for multiple tree species observed across the United States."

To be sure, a lot of the challenges facing the forests are challenges they've faced before. Die-offs are not unheard of, and are part of the natural cycle. But not on the scale that scientists now anticipate. Trees die in droughts, but they die faster when those droughts are accompanied by the kind of higher temperatures we've already seen and that scientists say are becoming increasingly likely. Even if droughts became no more common than they currently are, even if they lasted for only a comparatively short period of time, the combination, the NCA concludes, "could result in substantially greater mortality."

More dead trees means more risk of fire. "Projected climate changes suggest that western forests in the United States will be increasingly affected by large and intense fires that occur more frequently," the NCA report warns. And that means that whatever carbon is trapped in those trees while they live is released when they die. But even that pales in comparison to the other, more soul-wrenching and immediate consequences, which were hammered home in the summer of 2013, when nineteen members of an elite wildfire-fighting crew were killed when a savage blaze 85 miles northwest of Phoenix overtook them.[1] It was the largest loss of life in a wildfire in eighty years, and the worst loss of life among firefighters since 9/11. And the blaze was made devastating, many researchers argue, not just by the rising temperatures associated with the changing climate but also by increased development in supposedly pristine areas. By some estimates, more than 40 percent of the building permits issued in Washington, Oregon, and California in the past three decades have been issued in areas known as "wildland urban interfaces," rugged and often remote areas that were always prone to fire and are in danger of becoming more so.[2]

But perhaps just as important as his ability to cite the science, Tanner can cite the experiences of his fellow hunters and fishermen. To make the point, he tells about a trip he made a few years ago to Eureka, a remote little town in northern Montana. He went there with a friend who was lecturing on climate. "There might have been one guy in the audience who was younger than me," Tanner said. And so, perhaps out of respect, he asked the group a question. "'Most of you are not young guys, and I suspect some of you have been right here in Eureka for years and years. So you tell me . . . what changes have you seen over the course of your time here?'

"You know, it went on for twenty minutes. 'We used to get four feet of snow down on the flats, now we're lucky if we get six inches.' 'We never has this many dying trees before.' 'The snow used to come early enough that it pushed the elk down out of the high country before hunting season was over. We always had great elk hunting down here. Now the snows don't come until later, the elk don't get pushed down.' 'The ponds we used to hunt waterfowl on around here . . . a lot of those have actually dried up,'" they told him.

"They've seen all this stuff," he said. "But not one of them had ever made the connection to climate change."

That was, for him, a kind of epiphany.

"I came to the conclusion," he said, "that I needed to do something about this."

And so, a short time later, Tanner founded an organization called Conservation Hawks, a nonprofit, apolitical group of hunters and fishermen, designed primarily to educate other hunters about the risks of climate change.

It is, he argues, a moral responsibility for him. "I've got a son. He's eight. It's sort of an assumption of mine that . . . I'm going to be able to share the things I love with him as he gets older." But now, he fears that the changing climate threatens that assumption. "This is the one thing out there that can keep us from sharing our heritage and our traditions with our kids and our grandkids."

And that, he says, is the message he's been preaching to his fellow outdoorsmen. It's not been easy going, he tells me. Many of the hunters and fishermen he speaks to are—almost by default—conservative, and have for decades now absorbed the message of skepticism about climate that has infused the debate. One colleague even suggested to him that the whole idea of climate change was "a communist plot." And, echoing a complaint made by others who are trying to wrest the issue of climate away from skeptics on the right, Tanner says that most mainstream environmental organizations and philanthropic entities tend to favor predictably progressive groups at funding time, leaving groups like his to wither on the vine.

None of those obstacles have deterred him. Over the past few years, he says, "I've probably put 80 to 85 percent of my time into Conservation Hawks, creating the organization without ever getting paid for it, risking my marriage—my wife is not happy with the fact that the small amount of money I bring in has gone down. But it's one of the things I feel strongly about, and the people who have involved themselves with the organization feel so strongly about. We just figured we'd do what we can do now."

The message is, by necessity, broad, he says. "What we really want to accomplish right now is to get people to sit down and . . . deal with this. We're not tied to a particular solution. We're not tied to any particular ideology." The grand objective, he says, is to get the discussion started on ways to reduce carbon emissions to the point that the rising temperatures can be held to no more than 2 degrees Celsius, essentially where the broadest band of scientific agreement lies. The mechanism to do that is up to lawmakers and administrators. The cover they need to do it, he says, has to come not from the usual suspects who predictably fall on one side or the other of the cultural fractures, but from guys like the old men he met in Eureka or the thousands of others he's reached either personally or virtually since.

"Our approach . . . is that we never tell anybody 'this is what you should believe,'" he told me. "What we're always telling folks is that they need to believe what they see with their own eyes. . . . 'Does what you're seeing correspond with what the scientists are telling you?'"

He believes that sportsmen and -women are a resource that has been largely untapped in the discussions about climate. "Sportsmen come from a place of moral power when we do what's right for the resource, when we do what's right for future generations. Teddy Roosevelt saw that over a hundred years ago. If the people under Roosevelt hadn't put such emphasis on these things we would not have fish and game in this country to any great extent right now," he said. "I just look at this as essentially a moral issue."

If, he argues, the sporting community became energized and

active, it could have as great an impact on the climate debate as it has on the debate over guns in this country. "I mean, we have an NRA where nobody can ever possibly talk about taking away *any* of our guns—and I mean, I'm a huge gun supporter, but why don't we have the same thing when some piece of who we are is going to be removed from us forever? Why don't we stand up against that?"

Tanner admits he might be naive, but he believes that such a thing could happen. Hunters and anglers, after all, are people who hone their observational skills—success or failure depends on their ability to read almost imperceptible signs etched in the bark of a tree by an antler, or a broken twig where a broken twig shouldn't be, and they have learned from their very first hunt to alter their strategy according to what the woods tell them. They are also people who learn to trust their instincts. And often, when Todd Tanner talks about climate, that's what he's really talking about: trusting those instincts.

He does it with the grace of a storyteller. He did it with me, easing me into it, encouraging me to share a story or two about my hunting experiences in the East, bungling attempts that paid off more out of luck than any skill I might have had. And then he launched into his own hair-raising tale.

It was, he told me, a couple of years back, and he had gotten an assignment, the opportunity at the end of the season to fish along the Dean Channel, a deep saltwater fjord that cuts its way 75 miles inland into an all-but-forgotten corner of British Columbia. It's so remote that few anglers from outside ever find their way into it, and there are some years, he says, when not a single nonnative fisherman snaps a line over the water. It also has the second-largest population of grizzly bears in North America. The only way in is by floatplane, he told me, and as Tanner and a couple of his buddies, Billy and Dave, and the pilot, John, curved through the air over the wilderness, they spotted a large band of bears, twenty or twenty-five of them. From up there, it was hard to make out their specific features; you couldn't see the telltale hump that distinguishes a grizzly from a black bear, and so Tanner asked his host and pilot the obvious question: "How many of those are black bears and how many are grizzlies?"

"Oh," the pilot said nonchalantly, "those are all grizzlies. Black bears can't survive here. There's too many griz. They eat 'em."

Tanner made a mental note to try to catch a picture of at least one of the grizzlies—from ground level—which he figured would add a certain sense of majesty and excitement to his story.

For the next couple of days they holed up at night in the pilot's lodge a couple of dozen yards above the Dean, Tanner kicking back in the L. L. Bean deerskin slippers his wife had just bought him as a birthday present, enjoying the absolute solitude of a place so far off the grid that it could be reached only by air and contacted only by satellite phone. They spent their days fishing rivers and creeks even farther back in the bush.

And then one night, toward the end of the trip, as Tanner and his fellow anglers were clearing away the plates after dinner, the pilot came in to tell them that he had heard what he took to be a grizzly prowling around down by the floatplane. "If we go down to thc plane we can go around the bear and get some photos."

"It's probably too dark," Tanner replied. "But I'd love to see it anyway."

That was really only half true. Tanner did want to see the grizzly. There are few creatures on earth that are more awe-inspiring, even from a distance. But as he grabbed his camera and joined John, Billy, and Dave as they headed off along the two-track that snaked down to the fjord, plodding along in his deerskin slippers, something just didn't feel right. All of Tanner's instincts were telling him that this was a mistake. "I had a really bad feeling," Tanner told me. His instincts, he said, were sending him stark alarms. "I kept getting things like 'You shouldn't be here, you shouldn't be doing this,'" he said. "I literally talked myself into going forward, I was like . . . do you want to be the guy who says 'no, we need to go back to the lodge'?

"Two or three times I was actually on the verge of that," he said. But he resisted his instincts. Turned out he should have listened to them.

Because about a minute and a half after they left the lodge, without any of them noticing it, two grizzly cubs had circled around them, leaving the four men between the cubs and their very, very agitated mother. She was not a huge bear by British Columbia standards—maybe five hundred pounds—but she looked a helluva lot bigger when she reared up and started swiping the air.

Even then, Tanner assumed that the situation was under control, or at least that was what he told himself. "I was pretty sure she was going to get shot and I started jogging away, not wanting to be there when she got killed." His younger, slightly more athletic friend Bill was jogging alongside him. "And then Billy looked over his shoulder and started running really fast. That was a bad sign."

The old hunter's maxim "I don't have to outrun the bear—I just have to outrun you" flashed through his mind, and as he turned around, he could see that the bear, having already threatened Dave and John, was now charging after Tanner and Billy at full speed. Billy, like an acrobat, clambered up a nearby tree, the bear lunging after him, literally skinning the tree as she slid back down and turned on the pilot, rearing up and waving her claws at him, before turning on Tanner.

"You hear these stories about how time slows down? It really was like that," he told me. "I had this incredible clarity of thought, and the thought that was going through my head is that they were going to carve on my tombstone: 'He Tried to Run Away from a Grizzly Bear in His Slippers.'"

Tanner will never really know why he and his friends survived. Maybe she was confused. Maybe she was distracted. Maybe she just decided that she had made her point. In any case, the bear gathered her cubs and ambled off into the woods.

She certainly helped Tanner sharpen his point. And, he says, the point is simply this: He had failed to trust his instincts, and in so doing had made a series of small missteps that could have proved fatal. "The analogy to climate change is that it wasn't like I did one thing wrong and got in trouble for it," he said. "We made a whole bunch of what we thought were relatively innocuous decisions and . . . never [saw] that each one was taking us closer to this moment of ultimate peril.

"Climate," he told me, "is exactly the same way. We're making these small decisions now and we don't see the big picture. We don't see what could happen. What I'm deathly afraid of is that at some point in the not-too-distant future we're going to look back and say, 'My God, we're in huge trouble now! How did we get here? How is it possible that this is happening?'"

That's a message, he said, that resonates with hunters and fishermen. But only if it comes to them from a guy who is one of them. "If I go out in front of a sportsmen's club . . . to talk to these guys about climate change . . . wearing sandals and a tie-dyed T-shirt . . . what they're thinking is that I'm an environmentalist. I'm a green. Which automatically means that whatever message comes out of my mouth is highly suspect, if not something that should be ignored completely."

In the case of Conservation Hawks, he said, everybody who is associated with the organization is a sportsman or a sportswoman. "It's not like we're John Kerry going out there in the field with a goose gun. We're real. This is what we care about, what we truly love.

"Anybody who looks at us or talks to us knows . . . that we are in fact the real deal. And then we don't go out and frame things from a social perspective. What we say is that the thing that we care about is protecting our traditions and our sporting heritage—to be able to continue these things, continue to go out into the woods, continue to get out onto the water and have these experiences that are so much a part of our lives . . . to share these things with the people we love, with our kids and our grandkids . . . it's incumbent upon us to act as stewards and caretakers."

Little by little, he said, that message is getting through. A 2012 poll by the National Wildlife Federation, for instance, found that 66 percent of those who identified themselves as sportsmen believe the nation has a moral responsibility to address global climate change and 69 percent are behind the idea of reducing carbon dioxide emissions to do it.[3]

To be sure, there are times, Tanner acknowledged, when he loses hope, when he fears that changes in the environment are taking place too quickly, that the dice are already loaded against him, and that the divisions in political class are just too deep and wide to ever be bridged.

But the way he sees it, he has no choice.

"Nobody stands to lose more than sportsmen," he told me. "It's one of those things, if there was any way I could live with myself and *not* do this, I wouldn't do this. I'd do something a helluva lot more fun than write about climate change and focus on climate change. I don't want to do it. But then I go in and look at my boy sleeping at night. He does not look angelic during the day. But at night when he's asleep?"

He pauses for a moment. "We have an obligation."

{ SIX }

The Other White Meat

BY THE TIME THEY GOT THE BEAST TO THE EDGE of the loading dock at the old Nicholson Auction that night, he had long since lost whatever patience he might have had with the pair of gap-toothed, flannel-shirted drunks who had bought him, and the three-hundred-pound Yorkshire boar was now snorting and grunting wildly, pawing the manure-splattered cement with a murderous look in his little black porcine eyes.

My old man and I hadn't been expecting a floor show that night. We had driven the fifteen miles or so from the farm to the auction really just to get a sense of the market for heifers that fall. But that fell by the wayside as we huddled behind a whitewashed gate and watched the two drunks taking turns doing their pas de deux with Porky.

It had to have been the booze. Anybody who knows anything about livestock knows that a provoked pig can be the most deadly creature on a farm, a beast that can not only rip you to shreds with its razor-sharp incisors but also has no problem at all eating what's left. That was why huddling right beside us behind that gate were some of the most savvy livestock handlers in the neighborhood, among them Buddy Baldwin, our horse trader, an old rodeo rider who I'd personally seen stare down the most irascible horses with unshakable calm, a guy who just an instant before had easily subdued a singularly aggressive young Hereford bull. Buddy wanted no part of the pig. Neither did any of the other handlers.

But the two drunks were undeterred. I'm not sure when they decided to buy the pig in the first place. You would have thought that if they had planned the purchase in advance, they might have figured out a better way to transport the creature; they might have brought a truck or a trailer, at least. In any case, you would have thought they'd have found a vehicle more suited to the task than the Ford LTD convertible, the front seat littered with empty bottles and crushed Marlboro boxes, that they were driving that night. But since that was all they had, they decided that it was what they were going to use to get the pig home, and so they hit on a plan, of sorts. After goading and taunting the hog as far as the loading ramp, one of the two drunks staggered out into the parking lot and pulled the LTD around until it was parked about three feet below the edge of the loading ramp. I'm sure he took it as a testament to his residual sobriety that he remembered to put the top down.

The other one began gesticulating wildly, stomping his feet, making threatening faces, snorting back at the hog, and when that didn't work, he began grabbing clumps of manure from the ramp to toss at the beast, until finally, just as he had planned, the raging pig charged him, knocking both of them ass over teakettle into the backseat of the LTD, which screeched off into the darkness, spewing a trail of gravel behind it.

I have absolutely no idea whether the pig made it to his final destination. I don't even know if the two drunks made it home that night.

It's been more than forty years since then, but this still remains one of my most vivid memories, not just because of the utter dangerous absurdity of it, but because in a way, it's become a valuable life lesson for me. I see it as a metaphor for how foolish people can be when they're intoxicated with anything, be it booze or power or ideology, or any combination of the three.

Increasingly, these days, I find myself applying that metaphor to the way our elected representatives bungled the efforts to make climate change a legislative priority. Volumes have been written in recent years about how, over the past few decades, debate between extreme partisans in this country has become increasingly fractious and has now virtually become institutionalized. Congressional maps have been redrawn to the point that we now have 435 echo chambers in which politicians, egged on by talk radio and cable television pundits and bloggers on the margins, must often play to the most extreme minorities in their districts in order to make it through a primary season. The result is that more and more politicians find themselves in the same position as the two drunks that night in Nicholson. You see them at campaign stops and on C-SPAN, stomping their feet, tossing excrement when necessary in order to goad their rump constituents into charging headlong into their ideological LTDs, while the few cooler heads that have managed to survive in the Senate, Republicans and Democrats alike, try to keep a low profile. You can understand why. Just like the old horse trader Buddy Baldwin, they know better than to get in the way of a rampaging boar.

No event in recent years has more perfectly encapsulated that institutionalized dysfunction in Congress than the debacle that was the 2010 cap and trade bill. After much wrangling, the measure narrowly squeaked through in the House of Representatives with a slim Democratic majority and the hard-won support of a coalition of business bosses and honchos from several major environmental organizations. But then it died a slow and painful death in the Senate—though it had the support of most of the Democratic majority, it lacked the votes from either party to make it to the 60-vote threshold needed to avoid a filibuster.

That stunning defeat effectively ushered in a three-year deep freeze on any discussion of climate in Congress, an era of paralysis so pervasive that in June 2013, in what many environmentalists hailed as a milestone for action on climate and critics decried as a millstone around the neck of a struggling economy, President Obama tried to sidestep Congress altogether, unveiling a sweeping package of actions that the administration could take without congressional approval to reduce greenhouse gas emissions.[1] Among other things, the president's plan called for a reduction in the amount of carbon dioxide emissions produced by power plants by 17 percent from 2005 levels by 2020, called for increased fuel economy standards for vehicles, and directed federal agencies to get 20 percent of their electricity from renewables during the same period.

The president also curtailed U.S. funding—through the Export-Import Bank—for the construction of overseas coal-burning plants.

Even the most enthusiastic boosters of the administration's move, however, acknowledged that it was just a stopgap measure, and that in order to achieve meaningful long-term reductions in greenhouse gases and to shore up the nation's defenses against the ravages of rising temperatures, increasing sea levels, and increasingly volatile weather extremes, congressional action would be needed.

As Bob Ward, policy and communications director at the Grantham Research Institute on Climate Change and the Environment at the London School of Economics and Political Science, told the British newspaper the *Guardian* the day after Obama's speech, "Without the support of Congress for federal action, the United States will miss the opportunities that will be offered by the low-carbon industrial revolution."[2]

There was of course, no indication from Congress that that might happen.

The sweat hadn't yet evaporated from the president's brow before Republicans began attacking the plan as a job killer, and some Democrats in the Senate, like Louisiana's Mary Landrieu and Montana's Max Baucus, began distancing themselves from it.

In effect, the same fissures that doomed the cap and trade bill were still in play. If anything, they had become more pronounced.

There have been plenty of postmortems since the cap and trade bill's ignominious failure, and most focus on the role of special-interest money in killing the bill. But as Harvard political scientist Theda Skocpol wrote in the landmark 2013 report "Naming the Problem: What It Will Take to Counter Extremism and Engage Americans in the Fight against Global Warming,"[3] the measure was doomed from the beginning. Though Skocpol doesn't put it this way, she essentially argues that the initiative collapsed because both sides in the debate were trying to goad their supporters into the backseat of their respective LTD convertibles.

After the House version of the bill, sponsored by Democrats Henry Waxman and Ed Markey, squeaked through the House on a 219–212 vote, the far right, which has in the past two decades become increasingly extreme and increasingly powerful within the Republican Party, mounted a full-scale, well-funded attack on the measure, cashing in on the then-rising Tea Party movement, and scared off moderates on both sides of the aisle. "Oppositional lobbying and media campaigns went into overdrive and fierce grass-roots Tea Party protests broke out," Skocpol wrote. "During the summer Congressional recess, telegenic older white protestors carrying homemade signs appeared at normally sleepy 'town hall' sessions to harangue congressional Democrats. . . . Protests were bolstered by generously funded advertising campaigns targeted on senators who would be asked to decide about cap and trade bills in the fall."

The strategy paid off—for those who advocated inaction, at least. As Skocpol wrote, "In July 2010, Senate [Majority] Leader Harry Reid pulled the plug" after a series of potential compromises died on the vine, convincing him that he would be unable to meet the 60-vote threshold, and after more than a dozen members of the Democratic Caucus had in fact voted against using the 51-vote threshold to pass the bill.

"During this pivotal year," she wrote, "Republicans, including long-time supposed friends of the environmental movement like John McCain [who had initially supported cap and trade legislation] simply melted away; and in the end GOP senators unanimously refused to support any variant of cap and trade."

Hope evaporated, she said, in November when the voters toppled the Democratic majority in the House, turning the 112th

Congress into what she and other analysts have described as "one of the most right-wing in U.S. history."

But the blame for the bill's failure does not rest solely with the far-right wing of the Republican Party, Skocpol and other analysts argue. She sparked controversy with her assertion that leaders of major environmental groups failed to leave their trusty LTD convertibles in the parking lot as well, that they did not fully appreciate the volatility of the Tea Party or the impact its big backers would have on a public that was already cowed by the devastation of the Great Recession. "Like it or not," she wrote in an opinion piece for *Grist* magazine in 2013, "environmentalism has long been primarily a cause of the educated upper-middle class in the United States and it remains largely populated by experts and activists from that relatively privileged, non-majority class background. Needed global warming reforms . . . will require acceptance and some enthusiastic support from the majority of ordinary American workers and families. Almost all families now use carbon-intensive forms of energy to light and warm their homes. Because these families have not seen real wage increases in decades, they are extremely sensitive to even modest price increases in life necessities."[4]

In Skocpol's view, the whole debate played right into the hands of those who opposed carbon limits, allowing them, not entirely implausibly, to cast it as a showdown between hardworking regular Americans and the "elite." The advocates, whom she called the "climate reformers," never really rose to the challenge of explaining to those hardworking regular Americans what was in it for them.

Of course, as Dan Kahan of Yale has shown, and as Todd Tanner and Katharine Hayhoe have discovered, it's not just the message that matters, but who's delivering it. Those on the side of climate action might have had a better shot if they had had better spokespeople, folks who could appeal to those on the conservative side of the great divide as well as to those who, like most Americans, find themselves straddling our cultural fault lines.

Lately, however, a few increasingly powerful voices have begun to emerge on the conservative side of the political divide, people who are working to build support among conservatives and independents for efforts to regulate carbon emissions. Rob Sisson, a lifelong Michigan Republican, president and one of the founders of ConservAmerica, a group formerly known as Republicans for Environmental Protection, is one of them.

The way Sisson sees it, his job is to remind traditional "moderate, church-going Republicans who don't mind government protecting less-advantaged people and making sure the economy works well" that contrary to what they've been told by increasingly polarized media, climate change is a conservative issue.

To do that, he said, you have to counter the anger and fear that is the stock-in-trade of cable television and talk radio. He believes that talk radio and cable television—he blames Rush Limbaugh in particular—were instrumental in whipping up a false frenzy among conservatives that helped scuttle the cap and trade bill in 2010. "These are people who are frightened. . . . They think government let them down. They think the big banks let them down, and in fact, they look at energy companies today as the one shining hope to restore America's economy and pride."

They bristle, he argues, when affluent liberal environmental groups cast the energy industry as a band of mustache-twirling villains straight out of *There Will Be Blood*, and are far more likely to be moved to action by depictions of threats to game species and farm animals than by the pictures of polar bears that are routinely e-mailed to big liberal donors in places like San Francisco and New York. The result of that polarization is that "it becomes much easier for the fossil fuel industry to manipulate public opinion just like the tobacco industry did."

But Sisson believes that with the right kind of financial support, organizations like his can tap into that tranche of moderate Republican and independent voters and blunt the impact of the fossil fuel industry and the far right.

So far, however, the support has not been forthcoming. "We have begged and begged major funders, both individuals and institutions, to give us $250,000 so we can build our list. In 2010 we were begging for money to do it. Two hundred million was spent on the Waxman-Markey deal; we got like $15,000 of it," he said. And when it was done, a local congressman who had, to Sisson's surprise, voted against the measure and lost his seat anyway told Sisson that his vote would have been different had ConservAmerica and organizations like it been able to deliver even a comparative handful of Republican voters in support of the bill.

"We thought he was a yes," Sisson said. "He voted no. He called me and invited me to breakfast when he was in the district the next week. He said, 'Listen, Rush Limbaugh mentioned my name on the radio as someone who was getting ready to vote yes. So many calls came into my switchboard that my switchboard crashed for the first time in my career. . . . If [I'd] heard from just 100 Republicans in my district who . . . voted for me and who supported this, it would have made the difference.'

"I shared a copy of that quote with a lot of these foundations who turned us down for capacity-building grants," he said, "and they shot back, 'Calling us stupid doesn't help your cause.'"

But increasingly, he says, he's begun to hear rumblings within some of the leading environmental philanthropies that maybe he's onto something. Those rumblings haven't yet turned into the jingle of coins, he says, "but I've talked to a number of people [and] there's a growing awareness that it's an absolute necessity that a separate tranche of Republican voters be developed somewhere that can be mobilized for this sort of thing."

He's also hearing from Republicans and former Republicans who believe that the time is ripe for the GOP to take aggressive action on climate. Almost every day, he says, his organization gets a letter or a call or an e-mail from somebody who says, "I can't be a member of your group . . . because I'm not a Republican anymore. I didn't leave the party. The party left me."

Sometimes they tell him, "Thank God I found you because I was getting ready to leave the party."

Sisson sees it as his mission to provide a forum for those disgruntled and displaced Republican moderates, to mobilize them, to make sure that their voices are heard. And he is not alone.

Bob Inglis.

Perhaps the most high-profile conservative to take up the issue of climate change and to try to make it a distinctly conservative cause is former South Carolina representative Bob Inglis. During two stints in Congress, the first from 1993 to 1999 and the second from 2005 until his crushing defeat in the Tea Party uprising of 2010, Inglis admits he went from anti–climate change bomb thrower to crusader for what he believes is the most effective method of controlling carbon emissions, a revenue-neutral carbon tax coupled with a reduction in payroll taxes—a still-controversial measure that has nonetheless been embraced by such diverse characters as Al Gore and former Reagan economic advisor Art Laffer. Ideally, he sees it as essentially a dollar-for-dollar swap—that every dollar assessed on fossils would be offset by a reduction in payroll taxes.

He has no illusions about how polarized the political factions in Congress are over the issue, he told me. He understands it better than most, he said, "because I was one of the crew making it polarized."

For his first six years in office, Inglis was a dedicated climate skeptic. "I hadn't looked into it, but I just knew Al Gore was for it so I was against it."

To some degree, he argued, that kind of blind allegiance to partisan politics made a kind of perverse sense in a legislative body that was so deeply divided and where redistricting amplified the divisions by an order of magnitude.

"When I used to meet tour groups on the steps of the Capitol, I had a standard laugh line, and I'd say that representation in the House is based on population and we [South Carolina] have about 3.5 million people, so we get [seven] House members. . . . Delaware gets one. California has fifty-three. . . . If California got their act together they could run this place. Not to worry, it'll never happen, because they have both the *most* conservative and the *most* liberal members of Congress."

In fact, he said, the ideological balkanization had become so pervasive, so widespread, and so specifically targeted that quite often the most hard-core conservative districts and the most extreme liberal ones—districts that were separated by a cultural chasm as wide as the Grand Canyon—were separated on the map by just a matter of inches.

"Sometimes it was a matter of on that side of the street you could be a leftist and on the other side of the street you could be amazingly conservative," he said. "The more conservative the better."

The result, he argued, is that lawmakers have a big incentive to play hard to their base, taking positions sometimes even more extreme than the ones they actually hold. And it's not just true of conservatives, he said. As one reliably liberal New York congressman once put it to him, "I get up every morning and try to out-liberal my district and I've got to work hard to be at least as liberal as my constituents."

To be sure, he said, there were—and still are—harsh partisans in both houses, men like James Inhofe, perhaps the most vocal critic of the whole concept of human-driven climate change in the Senate, and John Shimkus, the Illinois member of Congress who famously argued that God would protect the United States from the ravages of climate change. "Some are true believers," Inglis said. "They are not likely to ever change their minds because they don't have the capacity to change their minds. They can't change their thinking because it is the only way they can think."

But those intractable voices, he believes, are the minority. The problem is that those are largely the voices that tend to get heard, especially among the party faithful and especially when it counts—in the two or three weeks leading up to an election. The vast majority of members of Congress are "reasonable, capable, smart," and, he added, scared to death of meeting the same fate that Inglis did in 2010 when he was crushed in the primary, getting 29 percent of the vote to his opponent's 71 percent.

To be fair, Inglis's road-to-Damascus conversion on the issue of climate change was not the only factor in his stunning primary defeat by Tea Party–backed candidate Trey Gowdy. Right-wing Republicans were every bit as angry over his support of the Emergency Economic Stabilization Act of 2008—the so-called bailout bill—and his opposition to the troop surge in Iraq that year.

But some voters saw his embrace of climate change as a betrayal of his conservative principles, and they were every bit as outraged by the way he came to his change of heart as by the change of heart itself, he says. It happened, he said, because of his son, who was just turning eighteen and was preparing to vote for the first time when Inglis decided to run for his old seat in 2004. "He said, 'I'll vote for you, Dad, but you need to clean up your act on the environment.'

"The thing was, his four sisters all agreed with him, my wife agreed with him, and I had this vital constituency. They could change the locks on the doors. I had to respond."

Unfortunately, as he learned on the campaign trail, a lot of his constituents found that tale less than heartwarming. "I thought it was a cute story and I told it all around, much to my peril, because it turned out that many conservatives were of the opinion that my kids should be listening to me, not me to them."

Underscoring that hostility among primary voters, he argued, is one basic emotion: "It's fear. Fear that the world is changing."

Many conservatives, he told me, have been conditioned by the media and by their elected officials, who are often themselves conditioned by the media, to see any action on climate "as an attack on a white, suburban way of life by elitists who want to order us to move downtown, walk to work in our Birkenstocks, eat tofu and live in some public housing hovel wearing gray uniforms," he said. "That's really what we see it as, and we don't want these clammy-hand bureaucrats telling us how to live our lives."

But, Inglis insisted, there is another aspect of conservatism that has yet to be tapped, and since his primary defeat, he's been working to tap it, traveling around the country, talking to small groups and large ones, pressing for what he contends is a classically conservative approach to climate change—his carbon tax plan. "What we're talking about is bedrock conservatism," he said. "We believe in free enterprise, we believe in the free market delivering efficient and effective outcomes. That's what conservatives believe."

There is, however, a built-in distortion in the system, he said. Direct subsidies to fossil fuel industries totaled more than half a trillion dollars in 2010, according to a study by Bloomberg. And while the industry may argue that on balance that's less per unit of energy than some renewables get, there's the hidden cost of policing fossil fuel operations, cleaning up after them when things go wrong, and mitigating the damage done by rising levels of carbon in the atmosphere. That money dwarfs the roughly $46 billion the government spent on renewables that year, and undermines the free-market principles that conservatives espouse, he argues.

But rather than push for a more level playing field, he contends, many conservative lawmakers continue to insist on subsidizing the fossil fuel industry, and they're encouraged to do that by the conservative press. "I find it really curious that the *Wall Street Journal* editorial page comes to the defense of coal and fossil fuels when they let tobacco and textiles [two industries that were deemed essential in Inglis's home state] go under the bus.

"I remember when I was first in Congress there was actually the thought that we could preserve the domestic textile industry through domestic content legislation. That was my charge as a new member of Congress . . . to try to save this industry. Well, the *Wall Street Journal* basically said, 'Get a better job. It stinks for you, but we're moving on.'

"Well, we did move on," he said. "A company called BMW—which stands for 'Bubba Makes Wheels'—came to Spartanburg, South Carolina, and we now have something like 6,000 jobs there and about 20,000 jobs in the supplier network. Thank you very much. We did move on."

It was a hard lesson, he said, but it was one that is elemental to conservatism.

And it's the message he is preaching as he travels around the country. It's a simple message, he says. "We stress accountability. That's a key concept to conservatives of all stripes. It's a key concept for social conservatives because people need to be seen as responsible moral actors. It's essential for economic conservatives because economic conservatives and libertarians are people who believe in accountable marketplaces . . . and you can't have functioning markets unless you have a cop on the beat and transparency and disclosure, and for national security conservatives, the fourth block of conservatism, you're looking for a way to break this addiction to oil and you realize the geopolitical problems that come with . . . making . . . oil a strategic commodity."

To be sure, it will take years for Inglis to build up the kind of grassroots support that he will need to prod his former colleagues in Congress up the manure-splattered loading ramp of public discourse to even listen to the kinds of proposals he's offering. He knows that. He has encountered opposition from fellow conservatives who claim that his approach would be "toxic" to the economy. He disagrees, of course, arguing that a carbon tax offset by a reduction in the payroll tax would help spur the economy. And little by little, he believes that message—a message delivered to conservatives by conservatives—is beginning to resonate.

"We conservatives have a very undeserved inferiority complex when it comes to energy and climate. We think we're no good. But actually we're very good. We have the answer the world is waiting for, which is to unleash the power of free enterprise."

When I hung up the phone after my conversation with Inglis, I found myself marveling at his sense of optimism, which I thought bordered on the naive. To clear my head, I hopped on my 50-miles-per-gallon motorcycle and rode north through the forest until I hit the quaint little nineteenth-century village of Honesdale. I've always loved that place. My family has deep roots there. My great-great-grandfather is buried in what remains of the foundation of the house he built. It's now the Catholic cemetery. His wife is buried beside him. So's one of my uncles, Billy, who died as an infant and for whom my own son, Liam, is named.

It's hard to imagine it now, looking at this sleepy little village near the banks of the Lackawaxen River, but once upon a time Honesdale was a center of commerce, the western terminus of the Delaware and Hudson Canal. As the mansions that line the main drag attest, a number of people got very wealthy in the early nineteenth century transporting raw materials from the Wyoming and Lackawanna Valleys, materials that were carried over the Moosic Mountains down the canal and across the river to Philadelphia and New York, where they would be fashioned into manufactured goods. And what were they carrying? Coal. Timber. Iron.

They may not have known it at the time, but the raw materials they carried for a tidy profit were the building blocks of the railroad, the very thing that would someday replace the canals. In fact, one of the first locomotives ever to operate in the United States, the British-built *Stourbridge Lion*, made its maiden run in Honesdale. In time, the Delaware and Hudson Canal morphed into the Delaware and Hudson Railroad. The canal had made itself obsolete, and it had made a profit doing it.

In essence, I realized, that was what Inglis was talking about. To be sure, the canal and the railroad that replaced it had a lot of help from government, but the transformation was a testament to the idea that we once had the ability to adapt to a changing world, and could use the combined power of government and industry to do it.

The question is, do we still have that ability?

Liam, between the ancient shell midden and the launchpad on the embattled east coast of Florida.

{ SEVEN }

Flying by Wire

WHEN WE FIRST SAW THE FARM, ONE OF THE things that drew my mother most to the place was the neat but whimsical geometric lines of the old white clapboard farmhouse, a sturdy peak-roofed main structure with an ambling front porch and three soaring gables in front and a two-story bow window that jutted out insolently on the east side of the house. It would have been perfect if the previous owners hadn't slapped together a ramshackle old barn-red coal shed and attached it—loosely—to the west side of the house.

Eventually, my father pledged, we'd get around to tearing the old thing down, maybe laying out a rose garden for my mother there. But in the years that followed, life intervened and the old shed, which got more rickety by the year, turned out to be a handy place to stash broken tools, snapped bridles, empty grain sacks, barn rope, and anything else that we didn't feel like carting off to the dump. Besides, it was also the perfect place to anchor the thin strand of electrified fence that we ran from the house to the shed, to some young saplings and fence posts before it encircled a paddock nearby. And so the old shed got a stay of execution.

At least until one particularly snowy winter did it in. Over the course of that winter, the accumulating weight of the snow gradually pried the shed from its moorings on the house, and half the ridiculous structure—the half not tethered to a sapling by a strand of wire—sagged a little bit more each day, until by spring it seemed to be groveling in supplication at the foot of the house. I'm not sure if it was that my father was tired of looking at the unsightly wreckage itself or that he couldn't stand to see anything—not even a rotting wood shed—posed so ignobly, but whatever the reason, it motivated him to at last deliver the coup de grâce to the thing, and early one spring morning he gathered up his pry bar, his wire cutters, his hammer, and me, marched us all to the shed, and set to work.

I don't know how long we were at it, furiously ripping and prying, tearing board from board, throwing each liberated piece into a pile in the middle of the yard, every one of them landing with a soul-satisfying *crack* or a less-than-thrilling *thunk*, depending on how far along it had been in nature's own slower but more sure process of decomposition, but before we knew it, the whole thing was gone—the torn tarpaper roof, the red-stained side walls, all except the westernmost wall. That was still there, leaning at a 45-degree angle, held up off the ground only by the gossamer strand of wire that was attached to the groaning sapling, four inches around, that was itself bent toward the ground at a 45-degree angle like a medieval catapult.

Now, if you've never done any kind of demolition work, there's something you need to understand right away: It's intoxicating. I don't know why. Maybe the act of destroying something with nothing more than your bare hands and some basic tools taps into some kind of primitive bloodlust, but whatever the reason, you become entirely consumed by the act, so much so that even a cautious, generally logical man like my father could be overwhelmed by it.

And apparently he was. After he spent God knows how long sweating and groaning trying to consign that shed to oblivion, I could tell by the look on his face that the last wall, defiantly dangling there in space, was an affront to him. It was an insult, a challenge that he had to beat back, and before I could say a word, he lunged for it, trying to get purchase on it so he could snip the last wire that held it up. But every time he tried to get his footing, the wall would teeter on its wire pivot, throwing my old man off balance and bucking him off onto the ground. It would then wriggle back and forth a few times giddily, as if it were mocking him. That's certainly the way he seemed to be taking it, and, standing a few yards away, removed from the fray, I could see that for my old man this had now become a matter of honor. It was going to be a duel to the death between my father and the wall. In a fury, my father jumped back from the wall like a boxer dodging an uppercut and, panting heavily, circled around it. Then he lunged for the sapling, shimmying halfway up the arch that the young ash tree had become under the weight of the wall. There was a look of pure, primal triumph in my father's eyes as he reached into his back pocket for the wire snippers.

"Umm, Dad?" I started hesitantly.

He just glared at me, as if I were taking the wall's side against him. And then he reached down with the wire cutters, pressing the wire deep into the *V* formed by their two sharp jaws, taking his time, savoring his imminent victory over the wall.

"But, ummm, Dad . . ."

It was too late.

I don't need to tell you what happened next. I'm sure there is some equation that can determine the precise force released when a four-inch sapling bent 45 degrees by the weight of three hundred pounds of rotting wood is suddenly released. Let's just say that for my father and me, the phrase "fly by wire" took on a whole new meaning that day.

I found myself thinking about that long-ago spring morning as I slowly made my way south on A1A, the low-lying seaside road that cuts through canyons of condos on Florida's increasingly densely populated northeast coast. That stretch of sand and marsh grass has always been vulnerable to sudden rages from the sea, from the days when the Timucua built their towering shell midden on the banks of the brackish lagoon called the Indian River. They were a people who had, historians tell us, eventually reached a kind of fatalistic accommodation with the sea. They would reap its bounty when it allowed them to, surrender everything to it when it changed its mind, and, when the sea permitted it, they would return and start all over again. It was stronger than they were. They knew it. They accepted it.

But then we came along. Europeans, morphing into Americans, people who—like my father with the wall—had an ingrained cultural proclivity to believe as a matter of perfect faith that there was always a way to master the forces of physics and nature, if only they were determined enough to find it. We tried to humble the ocean, to defy it, and, I thought as I drove south, we had done something the Timucua knew far better than to do. We had taunted it.

I had spent the morning in Jacksonville, a bright and shiny little bauble of a city that appropriately enough has become the insurance capital of the American South, and only by chance also houses the regional office of the Army Corps of Engineers, one of a small flotilla of federal agencies that among other responsibilities is tasked with the thankless job of trying to evaluate the risks of climate-driven sea changes and communicate those risks to people who, truth be told, really don't want to hear about it.

Matt Schrader, a young engineer with a surfer's tan and a mop of sun-streaked hair, had agreed to sit down over coffee with me. He explained that he wanted to talk to me not in his official capacity as a government employee, but simply as a guy with a point of view that had been forged both by science and by a lifetime spent living with the waves. I quickly understood why. Schrader's view of the challenges went far beyond the complex mechanics of coastal engineering that he's paid to assess. It touched on history, on philosophy, on the whole complicated web of factors that have made us more vulnerable to the rages of the sea.

We've always been, at best, tenuously perched on the unstable shore, he told me. And yet, we've always been drawn to it. The oldest city in the United States, St. Augustine, just a few miles down A1A from his office, is testament to that. The great midden at Canaveral, farther south, is testament to how far back into our past that desire to stake a claim alongside the power of the sea stretches. "Since the dawn of civilization, we've always lived near the coastline . . . for several reasons—it's a great place for food-gathering, for transportation of goods, for defense from your enemies, for attacking your enemies," Schrader said, and of course, over the past century in places like Florida, for snowbirds from the North to spend their twilight years.

But the coastline has never done more than simply tolerate us, and then only for a while. Storms, winds, the relentless action of the tide, would rise up in sudden anger and wipe everything away. Embedded in the history of the southeastern coast of the United States are the records of a culture that learned to embrace the violent rhythms of the sea. The Timucua, who trod lightly, knowing that what the sea gave, the sea would also take away; the Gullah/Geechee people, brought to these shores from West Africa in the eighteenth century so plantation owners could exploit their skills in raising indigo and rice in a harsh and threatening environment. Consigned to live in what the aristocratic planters then considered a coastal wasteland that stretched from St. Augustine to Wilmington, North Carolina, the Gullah/Geechee quickly learned to read the signs of the sea and patterned their lives in sync with it. They would build their homes out of materials that the sea provided, shells and sandstone, materials that often proved resistant to the relentless forces of the ocean, and they'd build them behind dunes or in protected coves. When things got bad, when a storm blew in, they'd use the lessons the sea taught them to adapt on the fly. I was reminded of a story an old Geechee man had told me once about how his father had saved his family's shell-and-stone house on Sapelo Island off the coast of Brunswick, Georgia, from a brutal hurricane that ripped ashore in the 1890s with such fury that it killed thirty-seven people in the comparatively advanced city of Brunswick. The house, the old man told me, had survived the lashing winds, estimated to have made landfall at well over 100 miles an hour. But the storm surge, several feet high, was another matter altogether. It threatened to lift the shell house off its foundation and dash it to pieces, and so the old man's father did a little engineering on the fly. With a deeply ingrained understanding of the dynamics of the sea, he ran outside and grabbed his ax—the same ax he used to dispatch the chickens for Sunday dinner and to split the firewood to cook them over—and calmly hacked two great *X*'s in the wooden floor of the house, to allow the water pressure to equalize. The cottage built for former slaves survived the storm, while a few miles away in Brunswick, the far sturdier structures built for former masters were washed away.

Those cultures are either gone, like the Timucua, Schrader told me, or, like the Gullah/Geechee people, threatened by the relentless pace of development, their legacy plowed under and paved over by the cultural descendants of the same people who, in an act of defiance against the sea, built their sturdy houses in Brunswick only to see them washed away.

Now, he says, there is an even more perilous threat boiling up out of the ocean.

Riding atop the crest of natural variations in the ocean's behavior, we now have to contend with rising sea levels and warming oceans, changes driven in no small part by our own behavior.

To be sure, the science behind predictions of climate change-related sea level variation is fast evolving. But it may not be changing as fast as the sea level itself. As recently as 2013, the Fifth Assessment Report of the Intergovernmental Panel of Climate Change (IPCC), basing its forecast on complex but still incomplete models, predicted that rising temperatures of between 0.1 and 0.2 degrees Celsius per decade, driven in no small part by the rise in the level of greenhouse gases in the atmosphere, will likely accelerate the rise of sea levels from the already alarming 1.7 millimeters a year that we have been experiencing for about a century to roughly 3.2 millimeters a year.[1] There are, climate scientists warned, a number of tightly interconnected reasons for the increase. For one thing, the rising temperatures mean that thousands of years' worth of water that had been trapped in glaciers and Arctic ice sheets is now slowly trickling into the earth's moving water table, from which it will eventually make its way to the sea—Greenland's melting ice alone is estimated to have added 0.59 millimeters a year to the rising seas between 2002 and 2011, up from 0.09 millimeters a decade earlier, and even in the Antarctic, where the ice cover is valiantly trying to defy reality and expanding in the east, the western ice shelf is melting.

What's more, that water itself is expanding with the temperatures. Even the most sober assessments and optimistic projections from the authors of the report estimated that global sea levels could rise by anywhere from 10.23 inches to 21.25 by 2100 if we were as a planet to put ourselves on a strict carbon diet by 2020. If we were to be a little less severe with ourselves, we might be looking at two feet, and if we were to continue churning out both carbon and consumers at the rate we have been, then we could be looking at three feet or more of water rushing at us from the oceans, more than enough to swamp the sources of 60 percent of the nation's gross domestic product, facilities that we somewhat injudiciously located in low-lying, potentially threatened coastal counties.

There were also unknowns that were then only dimly beginning to become apparent—questions like what would happen if we hit a tipping point, if the polar ice that used to reflect a significant portion of the sun's energy back into space was no longer there and both the dark water and the exposed land left in the wake of its disappearance began to soak up and retain that heat, further spurring the cycle. Then we could be looking at sea level rises of 20 feet or more. Though the risk was considered highly unlikely, someday we could vaporize all of the ice, turning Greenland back into the boreal wilderness it had been hundreds of thousands of years ago, and plunging much of the earth's surface beneath 200 feet of water.

Predictably, those who felt that we, or at least they, would be better off ignoring the warnings used the scientists' own measured language against them to justify inaction. The very fact that the scientists, a singularly cautious species, had offered a range of possibilities, outlined the points at which they were uncertain, and even gauged the possibility that any of the individual issues they had raised could be mistaken was proof to these critics that the science was incomplete, and that determination on their part gave them the opportunity to insinuate that at least some of the scientists were hyping the threat of rising sea levels. And, they noted, the scientists had been wrong before.

It turned out that the critics may have been half right. As the 2013 IPCC report noted, earlier projections had been off. In fact, they had been too low.

Several studies, including one released in 2013 by researchers from the Potsdam Institute for Climate Impact Research and Laboratoire d'Etudes en Géophysique et Océanographie Spatiales (LEGOS),[2] had found that indeed the overall warming trend for the previous twenty or so years had been about 0.16 degrees Celsius per decade, despite the fact that for a time temperature increases on land had sort of leveled off. That gain was pretty much what the IPCC had calculated in 2007. But it turns out that the sea level was rising far more in far less time than the researchers' still-developing models had led them to believe. In fact, sea levels were actually rising at a rate of about 3.2 millimeters a year, or about 60 percent faster than the IPCC's best estimate in 2007, the Potsdam researchers concluded, a finding that was reflected in the 2013 IPCC report.

There are still tremendous uncertainties about what all of this means. As Bob Dean, professor emeritus in the coastal and oceanographic engineering program at the University of Florida in Gainesville, put it, "The situation is that over the last century, global sea levels rose 1.7 millimeters [per year] . . . a little over half a foot per century. Now . . . they're measuring about 3.2 [millimeters a year]. So the question is, is that an oscillation that's going to come back to 1.7 or [is it] a change that will persist? And that we don't know."

The problem is that it's not strictly an academic question. It may not make a great deal of difference in places like the Maldives in the South Pacific whether the sea rises by three feet or six feet over the next century. Either way, that island nation would be washed off the map. But in a place like Florida, where the coast is crammed, like most coastal communities in the continental United States, with homes and condos and all the other accumulated weight of centuries of development, whether the rise in sea level is a foot and a half or three feet could make all the difference in the world. It would mean that entirely different and increasingly expensive strategies would be required to cope with the threat. And depending on how fast the level rose, strategies would have to change further, Dean said.

"I think this is one of the greatest questions our nation is going to confront—will have to confront," Dean told me. To be sure, there will be local variations. As in anything else in life, the risk will not be equally distributed, and each threatened community will have to make its own judgment on its own time scale—"Should they retreat? Should they take up a defensive position? Should they cart in enough sand to keep building up barrier beaches? And what about those places on the leeward side of the barriers? What will happen to them if the water rises one and a half feet? Or two? Or three?"

How long can they put it off? Most, perhaps, can wait a hundred years, Dean said. Some cannot. But for all of them, every possible response carries with it a whole host of interconnected risks, some of them economic, some environmental, some merely logistical, and some, at their heart, fundamentally moral.

In the face of the precisely imprecise prognostications of science, it's no wonder that some people, even people in positions of authority, would rather cling to the uncertainties, the caveats, and the qualifiers that the scientists dutifully include in their reports, so they can put off having to make a decision, a decision that regardless of what it might be would be certain to provoke angry howls from one side or the other on the climate debates.

In a way, they are all trying to keep their footing on a precariously dangling rotted wood wall, trying to determine where precisely they should make their stand when the time comes to snip the wire.

And that's what kept coming back to me as I sat in the gleaming atrium of the soaring insurance building that also houses the regional headquarters of the Army Corps of Engineers, chatting over a cup of fair trade coffee with Matt Schrader.

To him, the whole question pivots precariously on an issue that goes far beyond the dry calculations of engineering. "We can engineer anything," he told me, sounding for the first time in our chat like an engineer. "I mean, is it possible to so thoroughly engineer a barrier island that you could guarantee its safety," at least for the foreseeable future, assuming coast was no obstacle? "Yes."

The problem isn't whether we can do it, he says. "We have the science." The question is, at what cost? Like most agencies in government and beyond, the Army Corps uses three scenarios: one in which we do nothing and the sea level rises as it has, sending six feet or more of water surging toward our structures; a slightly more optimistic one in which we begin to grapple with the causes and only about a three-foot rise in sea level is expected; and a particularly optimistic scenario in which the sea rises only one foot.

In all three of those cases, there's a threat, he says, but the cost of responding to that threat is different in each scenario, and when advising city or state officials about the steps that would be necessary to gird their communities against potential losses resulting from sea level change in the future, the Corps urges them to consider all three scenarios and plan accordingly. There is a cold, calculating way to consider those options, Schrader says. It's a simple matter of balancing risk and benefit.

But those considerations go far beyond engineering. They are economic. But they're even more than that. They're moral questions, he says.

Take Dade County in Florida, for example, where more than 2.5 million people live less than 20 feet above the current sea level, and where not only do rising sea levels threaten to swamp them with storm surges, but the steadier intrusion of salt water into the already overburdened aquifer threatens their drinking water. Sure, there are mechanical fixes that could be employed—seawalls and desalinization plants could be built—but who would bear the cost? Or take Miami Beach, where streets designed to withstand floods once every five years are now inundated several days a year and where the water supplies are also in jeopardy, forcing those with foresight to consider what happens when people, including those with little income or on fixed incomes who are struggling now to pay $50 a month for water, find themselves in decades to come financing mechanical bulwarks to the water supply at a cost of $300 or $500 a month.

And that's before you even begin to consider the human cost of sea level change—the increased fury of storms, the havoc they wreak. Sure, there have always been hazards. "Once every ten years we have a major hurricane that comes up, kills thirty-five people and costs billions in rebuilding," Schrader said. There has always been an unspoken analysis that weighed those risks against benefits, against the economic and cultural value of the places we've built in harm's way.

But now, with the added threat that global climate change brings, those considerations have become even more fraught and even more complicated.

It's maddeningly complex. Later that afternoon, I sat on the patio outside my mother-in-law's house in twenty-first-century Cocoa. It's a bedroom community for the space industry that developed around a sleepy little village on the banks of the Indian River Lagoon. I found myself mulling just how precariously balanced we all are between the weight of our own handiwork and the straining natural and man-made forces that we might unleash by taking any step to address the problems we've created. And then I caught a glimpse of an image that gave me cause for hope.

You see, my mother-in-law had grown up on the banks of the lagoon long before the area was a playground for astronauts, back in the days when it was a citrus ranch town, a literal backwater. Though she has spent most of her life married to a rocket scientist—my father-in-law is an engineer who worked on several programs for NASA, including the Space Shuttle—she still retains most of the old values she was raised with, especially a deep and abiding Christian faith, which she expresses by scattering angels made of stone or plaster or porcelain throughout her home.

I had always regarded those little angels as quaint. They became a little less benign for me when I stumbled across a couple of statistics that, when pulled together, showed that while seven in ten Americans believed not only that the climate was changing and that we needed to take steps to both mitigate our contributions to it and to adapt to it, almost eight in ten Americans believed that angels are real and can affect our destiny. My first reaction was probably not very different from yours. I thought the data painted us as silly, superstitious people, recklessly indulging ourselves with the firm belief that some supernatural agent would—either by rapture or redemption—bail us out.

But as I sat there that afternoon, gazing past the stone angels on the patio toward the Kennedy Space Center and the looming Vehicle Assembly Building, one of the largest structures ever built by humans—and built on the shifting sands of an endangered coast as part of an unparalleled public project to propel us into space—I began to see the angels differently.

They were, I thought, not icons of some unhealthy connection to the past; they were instead an expression of hope, of a deep-seated belief that there is a future and there is something—in us or out there—that motivates us to go forward. In a way, I thought, that's the same impulse, spoken in a different language, that motivated the people who built the stone middens on these banks centuries ago; it's the same impulse that motivated that father to carve great *X*'s into the floor of the cottage in defiance of a hurricane more than a century ago, and it is the same impulse that motivated the men of my father-in-law's generation to build that massive VAB across the river so we could pierce the sky and peer into space.

The question was whether we could find a balance between the two. Could we harmonize them so the hope that the angels represented and the possibilities that the VAB represented could be aligned in such a way that we could honestly assess the challenges and begin to feel our way forward?

I didn't know the answer to the question. I still don't.

{ EIGHT }

Notes from the Ivory Clock Tower

THERE'S SOMETHING SOOTHING ABOUT THE MEA-sured, Teutonic, toymaker cadence of Dr. Klaus Keller's voice. It's calm but jovial. Listening to him as he holds forth in his office in the Deike Building at Pennsylvania State University, I can easily imagine him in another setting altogether, maybe a warm, candlelit, late-medieval workshop in some northern European city, all decked out in a leather apron, gigantic hand-polished magnifying glasses making his eyes look like saucers, peering down at a rough-hewn wooden tabletop, carefully molding each of a thousand intricate gears into one of those elaborate mechanical clocks that act out scenes from our collective psyche precisely on the hour from the bell tower of some grand public building.

In a way, that's exactly what he is doing. A noted scientist in his own right, Keller is now the university's point man on a five-year project involving thinkers from across the globe working to develop a way to fit together all the conflicting and often uncertain details that we know about the changing climate, to balance them against the vast amounts of information that we don't know with any degree of certainty, to factor in our fears and our hopes, and in the end, hopefully, develop a system to analyze the risks we face and weigh our actions.

His job, as he sees it, is to help create an intricate mechani-

Klaus Keller.

cal system that—once set in motion—would help us identify "the least stupid" responses to the challenges presented by our changing climate. To be sure, Keller's team is not the only group working to develop a method for evaluating the risks and responses to climate change. Similar efforts are taking place all across the world. And you can see their influence in initiatives like New York

City's $20 billion plan to armor the city against rising sea levels in the wake of the damage it sustained from Hurricane Sandy.[1]

In essence, Keller says, the team's objective is to explore a few deceptively simple questions that have maddeningly complex answers, if they have answers at all: "What are the risks due to climate change? How can we quantify them? Are they clear-cut or deeply uncertain?"

The truth is, he acknowledges, that for the moment a great deal of uncertainty about all of the questions swirls around climate change. Certainly, there are key areas where the vast majority of scientists generally agree: broadly put, that greenhouse gases are accumulating in the atmosphere at alarming rates, that those greenhouse gases increase—even if not in a linear way—the temperature of the earth, that those forces trigger certain predictable consequences like melting glaciers and polar ice, and that those in turn trigger sea level rise.

But when they begin to peer deeper, he says, the scientists find themselves debating about just how and just how soon the most severe of the consequences can be expected. "Sometimes . . . if you ask five experts, you get five answers. It's not that they're stupid," he says. "It's just that right now [we're] somewhat data-poor."

Unfortunately, he contends, given what we do know, we may not have the luxury of waiting until all the data are in. "It doesn't help to say we'll wait until we have a verifiable risk," he says. "That's not how it works, because we have to make decisions right now." That's not, he says, because some cataclysm is necessarily looming right over the horizon (though it might be—"there are wild cards"), but rather because we are a species that, in order to thrive, requires a complex infrastructure made of things that rust and rot over time. It's just a fact of life that we must constantly renew that infrastructure, piece by piece, each piece of it designed to last a few decades at most, and every time we do that we're making a bet that that piece of infrastructure will be able to function in the face of whatever changes take place in the environment.

On a practical level, he says, that means factoring what we

know about changing weather and sea patterns into designs that are meant to last twenty years or more. Say, for example, that you're planning to replace an aging pier at the port of Long Beach in California. You can't just do nothing, he says. The pier needs to be replaced. You have the option of rebuilding it exactly as it is and hoping against hope that the rising sea levels won't take it. But given what scientists have determined about sea level rise, that would probably be a sucker's bet, he argues.

"You're going to have more flooding. That's a fact."

But careening off in the other direction, adding tremendous cost to the project to build it far higher than the most realistic predictions of sea level rise would suggest to be necessary isn't much smarter, he argues. "Building it very high . . . five meters would be stupid. Doing nothing would be stupid. So what is the least stupid?"

What Keller is talking about is what climate geeks refer to as "adaptation"—an ongoing process of trying to deflect the consequences of climate change that we can reasonably expect within a specific time frame. By and large, these are responses that are narrowly confined to a particular time and place, a pier in Long Beach viewed in a twenty-year time frame or Lower Manhattan girding itself for what it can reasonably conclude might happen over the next fifty years.

But there is a far more complicated and, in a way, much more esoteric phalanx of questions that arise, he says, fraught with not just technical but moral problems, and that go further out in time than we are able to peer. "Risks from climate change arise over short time scales, like sea level change. But they also happen over longer time scales. We as humans, we live in the anthropocene, which means that humans are a key driver of the earth's systems." The carbon and other greenhouse gases that we've pumped—and continue to pump—into the atmosphere in our effort to become the planet's most dominant natural force, remain in play for millennia in some cases, and so the actions that we take now will have consequences that are destined to echo for a very long time.

"We emit CO_2, this gives us an increase in CO_2 concentration, this gives us climate change and climate change impacts," he says. "If you don't like the impacts of this logical chain, you must break the chain." Simply hunkering down and building up our defenses when we can no longer ignore the consequences is one response, but it has its limits. Radically altering the way we live, dramatically slashing our carbon output, heedless of the economic, social, and political consequences, investing in complex and risky geo-engineering projects designed to suck greenhouse gases out of the environment is at the opposite extreme, another response. Between those two points, however, are a thousand points, a thousand possible measured responses. And so the question is, how do you choose from the portfolio of those options?

What makes that choice so spectacularly difficult is the fact that as many unknowns as we encounter regarding the science itself, there is even more that we don't know about ourselves, he says. "The effects of our actions are [experienced] over centuries to millennia, but we have no strong . . . power to predict over centuries or millennia, or even [the wherewithal to] imagine how people will live.

"I have a son who is nine years old," he says. "I don't know how he will live in fifty years. He might live in a very different way. . . . It could be a great world or not. The point is, I cannot hide behind the fact that I cannot predict how he's going to live so that I make the world, maybe, a worse place."

The key, he says, is to develop a framework that allows people to appreciate the potential risks, to recognize ways that we can, as article 2 of the United Nations Framework Convention on Climate Change puts it, avoid "dangerous anthropogenic interference with the climate system," consider their costs and who will pay them, and make our decisions accordingly.

"This is now a much more well-posed question, because you do not need to predict specifically what Person X will do in 2075," Keller says. "You can start to think quantitatively about how to manage risks."

There is no single answer, of course, no silver bullet. Every step forward carries with it its own portfolio of risk; every potential strategy for mitigation is going to come at a cost, and somebody's ox is going to get gored.

But the way Keller sees it, his principal responsibility, both to his colleagues and to his son, is to develop a framework, a complex interlocking system of probabilities and possibilities, benefits and risks, that balances what we know with what we don't know, that as clearly as possible lays out whose ox is in harm's way. He's working to develop an architecture of what might be, and what could be, a logical, mechanical system of risk analysis so elegant that every quarter-hour or so, a cast-iron mechanical bear pirouettes out and strikes a bell with an iron hammer to issue a simple warning. Whatever you do, "don't break any important parts."

Of course, the problem is he's not doing this work in some pristine Alpine village of the mind. He's doing it in twenty-first-century America, a place where extreme partisans, certainly on the anti–climate change side, but to some extent among those seen as leaders in the movement to slow climate change as well, have turned the whole issue of climate into a cultural and political cudgel.

It's hard to look up and admire the intricate mechanical workings at the top of the ivory clock tower when the cobblestones you're standing on are littered with teeth and hair from the street fight that's going on right in front of you.

Just a couple of hundred yards from Keller's office, there's a guy who has been pretty well bloodied in that street fight. In fact, few people have been hammered as hard in the partisan debate over climate as Michael Mann, the influential Penn State climatologist behind one of the most iconic symbols of the climate debate, the "Hockey Stick."[2] The phrase refers to a graph that he developed back in 1999 that, using a variety of sources, including proxy data taken from tree rings and ice cores, showed a fairly stable trend for about 1,200 years until about 1990, when it shot up dramatically, giving the graph the appearance of a cross-section of a hockey stick.

That one graph, as much as anything, crystallized for a lot of people the grim mechanics of global climate change, and it established Mann's credentials as a force to be reckoned with in the climate debate. In 2007 he shared the Nobel Peace Prize with former vice president Al Gore and the Intergovernmental Panel on Climate Change in recognition of their work connecting climate change and human activity. He got an even more important award from his colleagues. They gave him a real hockey stick, signed by each of them, which now leans carelessly in a corner of his cluttered Penn State office.

But the hockey stick also turned out to be a pretty effective lightning rod, and to this day it, and Mann, continue to draw attacks, many of them bitter and personal, from a variety of opponents like the right-wing anti–climate change *National Review* (Mann is suing the magazine) or groups backed by the Koch brothers, who, in Mann's estimation, have a personal financial stake not only in discrediting his work linking climate change to human activity but also in preventing people like Klaus Keller from even establishing a mechanism to evaluate possible responses to the changes. Those attacks have continued apace. Two years after he received his piece of the Nobel Prize, Mann was drawn into the controversy over a series of e-mails stolen from the University of East Anglia in the United Kingdom that some climate change skeptics charged provided proof that Mann and his fellow scientists were cooking the books on climate, though for what reason, the skeptics never made clear. More than half a dozen investigations were launched, including one from Mann's own university, and all of them concluded that the skeptics' overheated allegations of academic fraud against Mann and his colleagues were baseless.

It's no surprise, therefore, that Mann, who has nothing but praise for the work that Keller and others like him are doing, and is even surprisingly optimistic that Keller's brand of elegant logic will eventually prevail, is also more acutely aware than most of the hazards posed by trying to apply such nuanced and textured reasoning in a country where the voices on either side, amplified by an increasingly partisan media, seem bent on exploiting our long-standing political and cultural fractures to advance their own agendas.

On the one hand, he says, you have those who insist on rejecting any notion that climate change is happening, let alone that it is likely to get worse. "There's the issue of whether climate change is real, and whether it represents a threat," he says. "There are objective facts, and facts are true whether you believe them or not." But he contends that those facts, which are subject to debate and interpretation, are in danger of being drowned in a puddle of doubt by people whose primary objective is to prevent guys like Keller from using the uncertainty as part of the equation to develop a series of responses to climate, and to instead use that doubt as an excuse to block any action at all. The sad truth, he says, is that deep-pocketed interests, groups like Americans for Prosperity, have been pumping tens of millions of dollars into the system, buying billboards and television ad time, supporting candidates who support doing nothing, ever, about climate, all in an effort to muddy the waters and protect their interests.

But the blame does not rest solely with the right, he says. "You have people on both sides of the spectrum who probably see the issue more emotionally than rationally. That includes both people who accept the science and those who don't."

While there is no parity between them, those who deny the reality of climate change tend to be better organized, better funded, and, in the political vernacular that increasingly defines this debate, "better able to stay on message," Mann says. He stresses that there are challenges for those on the other side as well, those who, in his words, "accept the science."

The danger posed by the increasingly quarrelsome tone of the debate is that as these cultural fractures widen into a chasm, important lessons from science can topple in. That was a lesson learned by Martin Hoerling, a comparatively little-known scientist studying the sudden drought that gripped the Midwest in 2012 after two years of extreme rain events. Hoerling and his colleagues released a paper suggesting that while rising temperatures linked to climate change probably played a role, the deciding factors may have been a collection of comparatively rare but natural events.[3]

There was an immediate response from fellow scientists, some of whom might have been a bit snarky but were generally respectful, disagreeing with Hoerling's assessment. Some, like Kevin E. Trenberth, head of the Climate Analysis Section at the National Center for Atmospheric Research and lead author of both the 2001 and the 2007 IPCC reports, argued that he had focused on too narrow a slice of the country, that he had overlooked the complicated interplay when the planet, even in the throes of record temperatures, tries to establish equilibrium, making one area drier while others get wetter, and that climate probably played a much greater role in the drought than Hoerling had assigned it.

No one, of course, had suggested that climate played no role.

But that's the way it seemed when the press and partisan bloggers got hold of the study. They broke along perfectly predictable lines, with anti–climate change voices, and those mainstream media outlets that were looking for the mythical balance, suggesting that Hoerling's study undermined the idea that global climate change was wreaking havoc on struggling farmers and those who relied on the commerce of the dwindling Mississippi River, while those on the other side, including the extreme voices of those who believed that any deviation from the orthodoxy that global warming was either solely or largely responsible for all unpleasant weather events, cast Hoerling's study as an act of heresy.

Making matters worse, Hoerling's report was issued just a few weeks after President Obama, in his 2013 State of the Union address, cited global climate change as a critical challenge, saying, "For the sake of our children and our future, we must do more to combat climate change."

The fact that a statement by a midlevel government scientist might be construed as a repudiation of the president's position (if you squinted real hard and looked at every other word) was grist for the slow-news-day mill, Hoerling complained.

"It was just a feeding frenzy," he later told me. On both sides, the reports claimed, "this report contradicted the president of the United States!"

Except it didn't, he said. "What the White House said was entirely true—there were these extreme events . . . extreme events happen. They will likely keep happening. Climate change is real, and climate change will also keep happening. . . . But they may not NECESSARILY be connected."

The real tragedy, Hoerling said, is that while opponents were ginning up a manufactured conflict between two ideologies, the real message of his report, a message that he and Trenberth could easily agree on, was utterly lost.

What motivated Hoerling to write the report in the first place was the fact that scientists like him and the people who depend on them for critical information were completely caught off guard by the Midwestern drought of 2012. They did not see it coming. And for Hoerling, the crucial question was why they didn't. In a fractured country where financial resources are often held hostage to political ideology, do we have in place the mechanisms to adequately predict what the combined forces of climate change and natural variation will do? Do we know how to tease out the signals of climate and better predict the weather, not at some distant point in time, but in time to plant next year's soybean or corn crop? "We wanted to be able to tell a farmer, we wanted to be able to tell the U.S. Department of Agriculture, hey, you're expecting 160 bushels of corn yield per acre? Based on what we might be able to see, you might want to scale that back to 123 bushels."

That's information that not only has value for the kind of long-term, big-picture planning that Keller is working on but also has real-world implications for the guy with mud on his boots who's out in a field somewhere in the middle of Illinois or Iowa or Kansas, trying to figure out how to hold on.

"This is one of the key research topics," Trenberth told me. "There's reason to think that we will be able to do better than we're doing right now. At the moment, the weather-prediction problem is largely an initial-value problem—that is to say, we track on a map where all the weather systems are and then we can predict where they are going and how they will intensify. There's skill for that out to about ten days. Maybe two weeks . . . but not a whole lot better."

And right now, given the economic and political realities of the country, the kind of short- and medium-term information that could tell a farmer when he should plant, what type of irrigation system he should configure, what crop to plant and what variety of that crop, is in short supply.

The sad truth, Hoerling said, is "if you were trying to anticipate the drought of May through August with the information you had in hand in early March and try to give advance warning before the farmers went into their fields . . . you would not have seen this coming. It was not in the cards."

As Hoerling put it, because we are in so many ways bumping up against the limits of our ability to predict the future, those farmers and others who are most directly affected by the vagaries of an atmosphere in flux, are all too often flying blind.

The thing is, though, at least some of them seem to be up to the challenge.

Illinois farmer Ethan Cox, in his workshop.

{ NINE }

"I Never Met a Liberal Before"

THE SUN WASN'T EVEN UP YET WHEN ETHAN COX tugged his work boots on, along with his old barn coat, the lighter one. He knew he wouldn't need the heavier one. He didn't even have to check the local forecast. It was going to be warm that day, low to mid-80s as the day wore on, he guessed, pretty much the same as it had been for quite a while. He glanced out the bedroom window at the sky. It was gray and brittle. It was going to be dry, too. That was no surprise either. The first week of March 2012 had been unusually dry. So had the whole month of February. In fact, the whole winter had been warm and dry. The yuppies and the liberals across the river in St. Louis or up in Chicago or out in San Francisco and New York all talked about that as being evidence that the climate was changing, that the bill was coming due for a century's worth of pouring all manner of poison into the atmosphere.

Ethan's neighbors thought that was kind of amusing. They saw the warm, dry weather as a godsend. After two years of record or near-record flooding, a deluge in 2011 so powerful that the Army Corps of Engineers decided to blow up the levees along the Mississippi River to keep Cairo, Illinois, from being washed off the map[1] and such brutal rainstorms a year earlier that the region suffered $3 billion in losses and crop and infrastructure damage that forced many farmers in the region to the brink of bankruptcy, to them the unseasonably warm and dry spring of 2012 was a sign from above that the worst was over, at least for now.

Ethan didn't think much of the liberals' point of view. They were always warning that something—the weather, the pesticides and fertilizers the farmers used, the very crops they grew, modified by biochemists in some corporate lab someplace—was going to tilt Earth on its axis and unleash all kinds of demonic forces. And it always seemed as if the only solution was to rein in farmers like Ethan, make them toe the line, regardless of what it cost in terms of productivity, regardless of what it cost the rest of the world in terms of slowing down the rate of food production even as the number of hungry mouths to feed skyrocketed around the globe. Not that he was entirely hostile to liberal ideas—he didn't mind the farm subsidies that came his way.

Ethan paused in the sleepy kitchen of the White Hall, Illinois, farm where he had been born sixty-five years earlier, poured himself a cup of coffee, and then trudged out the side door, across the yard toward the workshop, a kind of tractor shed and makeshift office that he had turned into the nerve center of the 3,000-acre corn and soy and cattle farm he had built the place into. He was moving a little slower these days. His knees weren't what they used to be. Neither was his heart. Seemed as if his body was every bit as creaky as the old corrugated metal sliding door to the workshop that grumbled and screeched in protest every time he hauled it open.

No, Ethan didn't think much of the liberals' point of view. But he didn't think much of his neighbors' unbounded optimism either. Maybe the liberals' warnings about global warming were overblown, but something was happening. Those two years of back-to-back storms were like nothing he had ever seen, and despite his best efforts to gird his land against nature's ravages—adopting no-till or strip-till farming to leave a protective cover on the ground and reduce the worst effects of erosion, for example—those storms had taken a toll, even on a farsighted farmer like him. His 2011 crop was a fraction of what it should have been. So was his 2010 harvest. Another year like that, and instead of getting paid, he'd owe money to the corporation that took his corn.

The thing was, there had been an ever-increasing number of years like that. In the fifty years since Ethan was a teenager, the number of extreme rain events—storms dumping more than three inches of rain on the sprawling farm fields of Illinois—had increased by 83 percent.[2] There were years like 1993 and 2008, years that saw the worst flooding in the Mississippi basin since the 1930s, and years like 2010 and 2011, when one after another, storms of amazing fury threatened to drown the young corn and soy before they got their heads up.

The good years were getting to be fewer and fewer. Ethan understood that. And as far forward as he could peer into the future, he saw that continuing.

He also understood in a way that most of his neighbors and even many scientists didn't yet that the volatility in the weather, those forces that were driving the rains, could—and no doubt would—just as easily shut off the tap altogether, leaving the same fields that only a year earlier had been inundated baking under a relentless, desiccating drought.

Those clear, warm blue skies that had raised his neighbors' hopes were, for Ethan, a bit more ominous. All winter long, it had been gnawing at him. Every time he'd head out on his ATV across snowless fields, he'd think back to those days six decades ago when he had been out here with his own father, plowing through axle-deep drifts in the first of several old Jeeps his father had bought—he had fallen in love with the things after a visit to Ethan's uncle in the mountains of New Mexico in the years after the Second World War, one of the few times Ethan had ever been that far from southern Illinois. That was back in 1954. The old man had figured that a Jeep like that would come in handy; they could use it to chase cattle or to haul back a deer after hunting, and it could even help them earn a couple of extra bucks if he fitted a blade to the front of it and hired himself out to clear his neighbors' lanes and driveways of snow. His father had been right. He usually was,

Ethan thought. Maybe that was part of the reason why Ethan still kept an old Jeep around the place, as a kind of rolling monument to his father's foresight.

Of course, Ethan hadn't really needed the Jeep much lately. The snows just weren't falling the way they used to. The cold didn't settle long enough or deep enough to freeze the water lines that snaked from the house his family had lived in for six generations to the livestock pens anymore. It seemed to Ethan that the deep cold and snows of his childhood were now as unusual as January thaws used to be.

Maybe it wasn't climate change, at least not the way the liberals talked about it. But something was changing—call it the weather if you like—and it had been changing for a long time. And there was no reason to believe that it wasn't going to continue. For how long? He didn't know for sure.

Ethan was a guy who measured time by the sort of work he did and when he did it, and by that reckoning, they hadn't experienced the kind of winters that were common when he was a kid, not in any of the years since he had sold a chunk of land to a corporate hog operation and leased it back, with the proviso that he not only would plant 800 acres of hay and 600 of corn on the land, but would also handle snowplowing operations for them for $75 an hour. There hadn't much snow to speak of since then. That was about fifteen years now.

That previous winter had been an especially mild one, and all winter long Ethan had been thinking about the lessons his father had taught him—how in those years when the real deep freezes and the snows didn't come, those years when the water lines never froze and they never had to haul water by hand to the hogs and cows, how those winters were, as often as not, followed by drought. Ethan's father didn't know the first thing about interdecadal variations in ocean temperatures, about how El Niño/La Niña cycles in the Pacific Ocean could cause flooding one year in the Mississippi River basin and drought the next. Hell, the old man didn't even believe that glaciers really existed. But he knew how to read the signs on his own land. And he taught his son how to do the same thing.

And all winter long, the signs were pointing toward drought.

The meteorologists who worked for NOAA didn't have any evidence to suggest that a drought was coming. The bean counters in St. Louis and Chicago and Rapid City, South Dakota—the folks who kept their eyes on such things and told farmers when they should plant—certainly hadn't seemed at all alarmed. Neither did Ethan's neighbors. After two destructive years, they were almost giddy about the odds for this year's crop. They would start planting by the end of March, or maybe early April at the latest, right when the bean counters told them they should. If they were worried about anything, it was the idea that maybe the storms were just laying low for a while, and that later in the year—May, June, maybe even into July—the savage rains would return. They thought that was unlikely, though. And as long as they could count on crop insurance, they'd probably be safe. But that meant they had to follow the rules. Plant too late, plant too early, and you ran the risk of not being able to get crop insurance. That was all right with them.

But it wasn't all right with Ethan. He was as sure as he had ever been about anything that the flooding rains weren't going to come. In fact, he was sure that pretty much no rain would come. In fact, he was so sure, he was willing to bet the farm on it. He wasn't at all certain that the farm could survive another year of bad crops, and there was even less of a chance that he'd be able to survive it if he didn't have crop insurance.

But a good yield could help him gain back some of the economic ground the last two years had washed out from under him. Especially if he could get his crops in early, before the other farmers got theirs to the elevators and while prices were still high.

The way he figured it, there was enough of the winter's scant moisture still trapped under the matting of last year's crops—matting left behind because he had adopted no-till, strip-till, and minimum-till techniques long before most of his neighbors did—

that his corn and his soy could get a head start, and be far enough along and strong enough to withstand the drought if and when it did hit. Of course, there was a major risk. If he was wrong, if there was no drought but instead those storms returned, with their flooding rains, their wicked winds, and hailstones the size of your fist, his lanky corn might be even more vulnerable than his neighbors'. He certainly would be more vulnerable, because he'd have little or no insurance.

Over the course of the winter, he sat in his workshop on an overstuffed couch—no longer presentable enough to remain in the old farmhouse where company might see it, but serviceable enough for the workshop—under the collection of arrowheads his father and grandfather had plowed up over the decades. Some of them were fake, but most were artifacts of people who had once lived on this land, including, perhaps, some of the same people who had built the massive ceremonial mounds at nearby Cahokia and who had vanished, possibly partly because they couldn't adapt to changes—call it what you will, the weather, the climate. Those challenges may not have been all that different from the challenges Ethan and his neighbors were facing.

Ethan sat on the couch and considered his options.

By the end of February, he had pretty much made up his mind. He called his twin grown daughters, Lydia Cox Hiesterman, newly married, living on a farm in Kansas with her husband and infant daughter, and Maria Cox, who was pursuing the life of a thoroughly modern professional woman in Rapid City. Both of them knew the risks. Not only had they grown up on the farm, they had both found their way into the crop insurance business. But they also knew their father, and though they certainly didn't see eye to eye with him on a lot of things—list any of the controversial issues of the day, from abortion to immigration—when it came to something like this, they believed with almost perfect faith that their father's instincts were worth more than all the carefully vetted analyses of all the bean counters in the world.

If their father was going to gamble everything on an early planting, they would stand behind him. Maria had one condition.

If he was going to do it, she was going to stand behind him not just figuratively but literally as well.

The truth was, she had been reconsidering her place in the world for a long time, and it had slowly started to dawn on her that all the things she had gone to Rapid City to find in herself—that sense of independence, the ability to make your own decisions and make your own way in the world, to trust your instincts, and reap the benefits or bear the consequences—those things weren't in the water in Rapid City. They were in her. It was what her father liked to call "the tilth gene," a peculiar sense of the connection between place and the people who live there and a sense that for either to be truly independent, the one is wholly dependent on the other. It had been planted in her by her father, just as it had been planted in him by his father, and by his father before him. If Ethan Cox was going to bet it all on his instincts, Maria Cox would be right there with him, making the gamble, too. She had a few changes to Ethan's plan, of course, but they could talk those over when she got there.

That pleased Ethan. It would be nice to have his daughter home. But first he had work to do. And so, on that unseasonably warm morning, March 13, 2012, weeks earlier than he had ever done it before, at precisely the moment that NOAA meteorologist Hoerling would later say no farmer could have gone out into the fields and predicted a drought, Ethan Cox fired up his tractor and headed out to his fields to plant. Because he was sure a drought was coming.

He was right. That summer, the rains were rare, and by the end of the season most of his neighbors, like farmers across the Midwest, saw their crops wither in the fields. Ethan, meanwhile, having listened to his land, harvested early and had one of his best crops in years.

THE WORD THE SCIENTISTS USE IS "ADAPTATION," and for a very long time it was considered a dirty word among many of those involved in the environmental movement. In his

1992 book, *Earth in the Balance*,[3] Al Gore warned that any attempt to adapt, to try to make peace with a changing climate, was in effect a capitulation. By 2013, faced with rising sea levels and with the prospect of more furious storms, with more vicious prolonged droughts, like the one that has settled over much of Texas for years, or, as in the case of the Midwest drought of 2012, with droughts that appear suddenly, Gore had changed his mind. "I was wrong," he stated bluntly in his 2013 book, *The Future: Six Drivers of Social Change*.[4] Adaptation and mitigation, efforts to reduce the amounts of greenhouse gases pumped into the atmosphere, had to work hand in hand. By 2013 that message was beginning to be heard. Sure, there were still those who resisted, extreme voices on the far right, people like Oklahoma senator James Inhofe who insisted that the whole idea of human-driven climate change was a hoax perpetrated on a gullible populace as a kind of stalking horse for a left-wing plot to turn the whole country into a dystopian, socialistic hellscape like Cuba or, worse, like Ithaca, New York. But increasingly, and on a very local level, places that were already starting to see the devastating cost of inaction, places like New York City in the wake of Sandy, South Florida, and the coast of California, were beginning to realize that in this polarized nation, action from above was unlikely, at least in the near term, and that they needed to adopt measures that recognized the changing reality—to adapt.

In other words, they were starting to realize what Ethan Cox and thousands of other farmers and ranchers all across the country had understood for a very long time: It was up to them.

As I've traveled around the country these past few years, I've seen it again and again, that deep, in-your-bones understanding that things are changing, carved into the brows of farmers and ranchers and fishermen. But there was something else there as well—a sense of responsibility, a belief that if they work hard enough, farm smart enough, have enough faith in themselves and their abilities that have been handed down from generation to generation, they can survive.

Scratch any of them and you'll as likely as not find a climate skeptic. These are, after all, conservative people, by and large, and the issue of climate has become a cultural touchstone, a defining dogma that fits neatly into the whole catechism of both the right and the left and occupies a space somewhere between gay marriage and gun control.

But probe a little deeper and what you find is that fundamental sense of pragmatism mixed with self-reliance that has always been a part of the character of rural Americans. A lot of them, like Ethan, are facing a problem that shows no sign of improving on its own. And so they believe it's up to them to take steps to plan for the future. There are fancy words the academics use to describe those steps: "Mitigation." "Adaptation." A lot of rural Americans just call it farming.

I saw that in the spring of 2013 when I wandered onto David Ford's 7,000-acre cattle, corn, wheat, and cotton operation on a sun-bleached stretch of back road outside the little Panhandle town of Dumas, Texas. Just two days earlier, I had been sitting in the College Station office of the Texas state climatologist, John Nielsen-Gammon, a man who has the unenviable task of having to consistently deliver bad news to farmers and ranchers like Ford, and I asked him for a prediction. He told me that the merciless drought that had been hammering the Panhandle for years, punctuated only by sporadic brief and often violent rain, hail, and snowstorms, was likely to last as far into the future as he could peer. Part of it was natural, he told me. Part of it, perhaps 10 or 20 percent, was the perfectly predictable result of adding roughly 1 degree Celsius to the temperature in that region through our profligate use of fossil fuels and our release of other greenhouse gases, actions that would continue to alter the climate for centuries. This was the new normal, he said. I told him I was heading up to the Panhandle and asked him what he would advise a guy like David Ford to do.

"I generally assume that they're in a much better position to evaluate all of the considerations than I am," Nielsen-Gammon told me. "So I don't tell people what to do."

And that was just fine with David Ford. As he leaned against his gigantic Case 500 tractor, a 600-horsepower air-conditioned, computer-guided monster, and surveyed the enormous spread his family had amassed over the past three generations, Ford told me he doesn't know whether he believes in man-made global warming. "You get into this global warming," he said, "everybody gets pretty hot over that. I don't know that I agree with global warming." But he does believe in the drought. And like Nielsen-Gammon, he doesn't see an end to it. More importantly, he told me, he believes in his sons, both grown men who have returned to the family farm, and in himself, and their ability to battle the weather, at least for now.

As the drought has persisted, he's come to understand that a deeply held belief was deeply flawed. "We had what we thought was an unlimited amount of water," he said. The drought taught them that they didn't. "We thought it had to rain sooner or later. No, it didn't."

And so, he said, he and his sons have changed the way they work the land to adapt to the new reality. Like Ethan Cox in Illinois, they've aggressively adopted a policy of doing as little plowing as possible to get their seeds into the ground. Not only does it help preserve moisture during those protracted periods when there isn't a drop of rain—he got less than an inch from January to the end of May 2013, meaning he can draw less from his already overburdened aquifer—it also protects the soil when those sudden quickly vanishing gully washers do occur. "We leave this mulch," he told me, pointing to a field littered with last year's corn stubble. "If we get any moisture, it'll get underneath that. That helps shade the ground and keeps the moisture there longer. If we get a big rain, in my fields there [isn't] any water running out of them because the organic matter left in the field is holding the water." He spent the first few years designing and perfecting a rig to handle the ground precisely the way he wanted it to. It's an ongoing experiment, he said.

They've experimented with different kinds of crops, different kinds of seeds, including the genetically modified drought-tolerant seeds made by a handful of multinational companies—

Monsanto, Pioneer, a company for which Ford tests and sells seeds, and others—products that, he notes with some degree of amusement, raise the ire of many of the same people who argue most passionately for the need to do something immediately to fend off the hazards of a changing climate. And they've experimented with different techniques to reduce water consumption, discovering in the process that corn planted earlier in the year seemed to use about four inches less water per acre than corn planted later. For now, that experiment is limited to a few test rows at the edge of Ford's farm. He's tried to interest scientists from the state and its universities in expanding the experiment, but so far he's had no takers. He's at a loss to explain why.

It doesn't matter. He and his sons have other options for trying to adapt. In recent years that effort has included changing the mix of crops that they plant, quadrupling the amount of cotton they produce—their yield went from 300 to 1,200 pounds in less than four years, because cotton fares better in the relentlessly arid environment of the Texas Panhandle than corn or wheat, he says. Besides, they don't need as much corn anymore. Unlike other areas of the country where much of the corn crop goes to the production of ethanol, most of Ford's harvest—98 percent of it, he estimates—used to go to feed the vast herds of cattle that used to roam the region. Most of those cows are gone, sold off or moved out of state, he says, victims of the drought. He's kept his own herd together, in part by culling out the temperamental, high-impact European breeds he once ran—Charolais, Salers—in favor of more rugged, smaller frame, and less voracious Angus crosses, and he feeds them not just the corn he grows, but the residue left over when his cotton is ginned. To be sure, Ford is aware that there is a raging debate in some quarters over whether operations like his are the most effective use of farmland, whether we might be better served in the long run by evolving to the point where rather than growing food to feed cattle, we'd raise crops that people could eat, but for the moment, he says, that's a debate for academics. Adaptation, he says, doesn't mean just adapting to the changing realities of rainfall and temperature; it also means adapting to the economic realities, which, if they ever change, will change glacially.

What's most intriguing about David Ford's experiments in adapting to a changing environment is that they also are driving a bit of what scientists like to call mitigation as well. The experiments he's conducting demand less diesel fuel, he told me. He now makes about a third as many passes on a field as he did before he adopted minimum-till practices, and he spends two-thirds fewer hours in the tractor seat over the course of a year. At the same time, however, he requires more-precise and more-powerful machines, hence that massive beast of a tractor that he climbed down from when I first showed up on his land. He had guided me by cell phone as I made my way to the back field in my rented ecobox (a car that got about 39 miles to the gallon), and he got a bit frustrated with me when I kept getting lost, making a left at the corral when I should have made a right, or heading straight past the tractor shed when I should have looped around it.

I understood his frustration when he showed me the machine that he had just been driving. It was outfitted with a GPS-guided auto-steering system so utterly precise that it could travel the length of his longest field and never deviate more than an inch in a half mile or more. Though it churned out about the same bone-crushing levels of horsepower as a NASCAR racer, the computer-managed tractor seldom crept above an idle. That was a little disconcerting at first for a guy like Ford who had grown up jamming gears on old-fashioned tractors. "I'm used to the old school," he told me, "the old tractors shuddering and pulling, that's the way she ought to run, and all of a sudden I get in there, I watch the RPMs drop 400 and I go up three gears from where I was. And I'm like, this does not make sense. I don't hear the engine roaring. But yet my fuel consumption is down."

In fact, even fitted with a Diesel Exhaust Fluid system, which uses synthetic urea to reduce noxious emissions from the tractor's exhaust pipe, the lumbering machine burns less fuel in a month than he used to burn in a week in the old days.

To be sure, the amount of fuel he's saving is a drop in an ever-rising ocean of fossil fuels that we continue to consume at an increasing pace, but in some small way, the experiments that David Ford and his sons are running are evidence that maybe Al Gore was right in 2013 when he said that he had been wrong in 1992, that maybe adaptation and mitigation didn't have to be at odds with each other. Maybe they could complement each other, and even if you didn't necessarily believe that humans were altering the climate, maybe there were good economic reasons to continue the experiment. At least that's the way Ford sees it.

Of course, not all the strategies to deal with the changes that rising temperatures have wrought on the Panhandle are as encouraging as the ones Ford has employed. You can see that when you pull into the little town of Guthrie, 50 miles south of Dumas, across some of the most drought-stricken land in all of Texas. If you had driven through these parched lands just a few years ago, you would have seen something very different. You would have seen thousands upon thousands of black baldies and other breeds of beef cattle grazing in pastureland that was never the greenest on earth but was enough to support a massive cattle industry. You would have seen water tanks brimming and ranch hands for massive operations like the historic 6666 Ranch, the foundation upon which the little town of Guthrie was built, tending it all.

It had never been easy to tend thousands of cows on just under 300,000 acres split between the two ranges that the 140-year-old outfit runs, one in Guthrie and the other in Central Texas, said Joe Leathers, manager of the 6666 cattle operation. There was always competition for water between the ranchers and the others who depended on the limited rainfall and struggling aquifers, among them the burgeoning natural gas industry, though for the most part they were able to accommodate each other and both were able to thrive.

At least as long as the rain kept falling. But then the rain stopped falling. It was the summer of 2011 when things started to get really bad, Leathers told me. "We went into . . . summer as dry as we've been in years, with no predicted rain." In a lifetime spent ranching in Texas, Leathers had seen plenty of droughts, and at first he and the other ranchers assumed that it would pass. After all, droughts had always passed before. "We did the typical ranch deal during drought situations. We culled our cows. Got rid of our old cows, weaned our calves early, and lightened up."

For the next several months, he said, he and the nearby ranchers bided their time, hoping that things would change. "We were kind of kicking the can down the road two months at a time, thinking it's gonna rain." But it didn't rain. "Our earth tanks started drying up," he said. "We were down to windmills and pipelines."

In Austin, in Washington, in the think tanks in the East and the universities, there were a lot of people who had the luxury of debating whether climate change was real, whether it was man-made, whether we could do anything about it, or whether we should.

No one in Guthrie had that luxury. It wasn't raining. It wasn't going to rain, at least not anytime soon, and so Leathers and some of the other ranchers got together and faced some hard facts. They took a survey of their water supplies, factored that out over three quarters, and then decided that they'd have to cut their herds in half.

But even that turned out to be too little, too late, and at last, faced with the prospect of a drought that could go on for years, Leathers and a few other ranchers decided to do the unthinkable. For the first time in 140 years, 6666 Ranch would move its herd, virtually all of it, out of Texas. "By the end of that summer I had moved approximately 4,000 cows and had eight different places leased from Nebraska and South Dakota to Montana."

Joe Leathers would never put it this way, but in a sense, the 6666 cattle operation had become a salvage operation, one of the nation's first climate refugees, perhaps—among the first to adopt the most daunting of adaptation strategies. Retreat.

It may be hard to believe, but that strategy has, at its core, a fundamentally hopeful message. And it also contains a hefty dose of respect for the power of nature and for the responsibility to tend it. Joe Leathers wouldn't use those words, though. Not exactly.

Instead, he used these words: "The goal is to stay there, and hopefully when it rains here, we can restock.

"The big deal here in Texas for us ranchers—and I can't speak for anybody else in Texas—but for us ranchers, you have to take care of your country. The worst thing you can do in my opinion is overstock your country and feed it off down to the dirt just trying to hang on. So we're trying to leave the country in good enough shape that when it does rain we'll be back in business pretty quick."

And tending to that country, for a guy like Leathers, also means tending to the people who tend it. In 2013, more than a year after Leathers moved the herd out of Texas, 6666 was continuing to pay its ranch hands, some of whom have been with the outfit for three generations. They were spending most of their time mending fences or hacking back the drought-loving brush that threatens to swallow more of the range with each passing rainless day, and periodically, he'd round up some of them—young Steve Briggs and Steve's father-in-law and a few other guys—and head north to tend the herd for a few weeks. For as long as Leathers can manage, they'll still collect a paycheck, and that paycheck will help support the little town of Guthrie, Texas. Because someday, he believes, it will rain.

If you ask Joe Leathers whether he believes in man-made global climate change, he'll tell you no in no uncertain terms. "We have faced droughts and severe rain and severe weather conditions in the United States and globally since the beginning of time. . . . I think the worst thing anybody can do is think that this is some sort of man-made problem and get men to try to fix it."

But ask him what his responsibility is and you get a very different answer. "What we've got to do in agriculture is the same thing that we've done my whole life and my forefathers before us—learn to live with nature and work with nature."

In essence that's the same answer that Ethan Cox had given me when I turned up on his front porch to talk to him about his all-or-nothing-at-all gamble against the drought. He had, of course, been right.

Cox admitted that at first he had been reluctant to meet with me. "I never met a liberal before," he said, and so, he confessed, he had his daughter Maria, who was, in his estimation, a bit worldlier than he, vet me first, to make sure that I wasn't some tie-dyed extremist.

I had already known that, of course.

He needn't have worried, I told him. Though there was a time in my youth when I strutted around with a copy of Mao's Little Red Book peeking out of the back pocket of my jeans, these days all that was likely to be stuffed in there was a pack of Rolaids. At this stage in my life I'm a lot less interested in stoking the fire in my belly than I am in simply finding a way to manage it.

Besides, I wasn't there to lecture him. I was there so he could lecture me, explain to me how it was that he had seen the devastating drought coming when no one else, not even the scientists paid to warn him, did. I was there so he could explain to me how it was that even if he didn't believe that we were responsible for at least a portion of the changes, he still had the foresight to armor himself against them.

He turned to his daughter and smiled. "You're gonna think your dad's a flaming northeastern liberal," he began. She smiled back. "There's two types of farmers," he said. "One either has the tire gene or the tilth gene. You either want a big shiny red or green tractor to disc up all the ground or you want to conserve that ground and make sure you have the right tilth.

"I'd like to say that I have the tilth gene rather than the tire gene."

We had been sitting there, sipping coffee and talking, for more than two hours, Ethan hunching forward every time I reached into my tobacco pouch and rolled a cigarette. Knowing about his health problems, I had offered not to smoke, at least not inside the workshop, but he would have none of it. His doctor may have ordered him to give it up a few years earlier after his heart had started giving him trouble, and he had reluctantly kicked the habit, his only real vice, but he still enjoyed smoking vicariously. Besides, he hadn't seen anybody roll one in decades, he told me, and it sort of took him back to his youth.

We had, by that point, covered everything I had come there to talk with him about: his prescience about the drought, his strategies to combat it, the risks he had taken, the way his approach to farming had changed and adapted with the times and with the increasingly volatile weather. That's when Ethan turned the tables on me. He had not been kidding when he said that he had never met a liberal before.

Where would he have met one? He had, he told me, never been on an airplane, never taken his kids on vacation—the cows wouldn't permit it—and rarely ventured more than a few miles from the little house where he had been born or off the prairies that stretched out forever from the little village of White Hall. He was part of the place, and the place was one of those places where things change slowly, if at all. I had recognized that earlier that Sunday morning when I stopped at a little convenience store in White Hall to grab a cup of coffee and to phone the Coxes to tell them that I was a few minutes away. I could have been in any farm town in America, sitting there at a chipped Formica table, listening over my shoulder as the slightly overweight middle-aged guy behind me talked about the local bars he had visited the night before in celebration of his fiftieth birthday. His friends called him "Cheetah," a nickname he had picked up in high school because, though you wouldn't know it to look at him now, back then Cheetah could run like the wind.

It's hard, I'm sure, for a lot of Americans, a nomadic people who for the better part of the last six decades have moved from identical suburb to identical suburb, to imagine that there are still places where a guy who's eligible for his first AARP card is still called by his high school nickname, and is still called that by the guys who gave it to him. But there are. And there are millions of people out there who live not far from their fathers' graves in places like Dumas, Texas, and Ellsworth Hill, Pennsylvania, places where the rules of the game change slowly, if at all, and where when change does come, it is often viewed with suspicion. These are places where individuality is prized, in large part because there is a social support structure that has been tested over generations, and certain key values are so much a part of that social fabric that there's no need to talk about them at all except to use them as a marker to draw the boundary line between who "we" are and who "they" are out there in the suburbs and the cities.

It would be easy to dismiss all that as quaint and provincial, and too often in our political and cultural discourse, we do just that, with the right trotting out cartoon images of good, hardworking rural Americans to bolster some political argument, and the left using equally cartoonish images to depict them at best as rubes and at worst as ignorant cogs in a vast machine who are working against their own best interests.

That, of course, presupposes that people in rural America are willing to accept, hook, line, and sinker, whatever some guy with polished shoes tells them, and anybody who's spent any time in rural America can tell you that is certainly not true.

It was precisely because Ethan understood that he couldn't believe what he had been told that he began quizzing me, not on my attitudes about global climate change but on a series of issues that, at least on the surface, had nothing to do with the subject. We talked about abortion, a subject that he sees as an indication of social decline, and he understood when I told him that I was pro-choice but that didn't mean that I was pro-abortion. We talked about people we knew who had struggled with the decision, we put faces on the issue, and talked about our shared responsibility to those people we knew, and to those we didn't. The subject changed to guns. We talked about the semiautomatic rifle with the 30-round clip that he keeps in the back of his Jeep, on the off chance that he has to plink a coyote before it gets a calf (he seldom hits one), and he allows that if it would stop the kinds of massacres that had taken place a few months earlier at Sandy Hook Elementary School in Connecticut, or a few months before that at a movie theater in Colorado, he'd gladly swap his .226

for the flintlock rifle I use to hunt deer in Pennsylvania. He even chuckled when I told him that I carry the gun that the Second Amendment more or less explicitly permits me to carry.

For the next two hours or so, Ethan and I talked. It wasn't an interview anymore. It was a conversation between two old men who, while we may come from different ideological camps, have each managed in our lives to cheat catastrophe long enough to learn to listen to each other.

And at the end of the conversation, I rolled a final cigarette, and Ethan took a deep breath when I lit it. "You know what, Ethan?" I said. "We've just sat here for the better part of four and half hours, a good old-fashioned rock-ribbed conservative like you and a good old-fashioned dyed-in-the-wool liberal like me, and we've touched on most of the major hot-button issues in the culture wars—abortion, same-sex marriage, guns, even climate. On about 85 percent of those issues, you and I could find enough common ground to find a shared purpose. On another 10 percent or so, we could at least reach an understanding. There was maybe about 5 percent where the differences were just too great. But we could set those aside, at least for now."

He agreed.

"So why is it," I asked, "that when I hear people talking about you, and you hear people talking about me, the only thing they ever talk about is that 5 percent?"

Certainly we need to heed the warnings of scientists who tell us that the evidence is mounting, the climate is in flux, and there is a profound risk that if we don't both armor ourselves against those changes and work diligently to reduce whatever part of it we are responsible for, the consequences could be extreme.

It is also true that there are profound lessons that must be learned from folks like Ethan Cox and Joe Leathers and David Ford, lessons about how to adapt to a changing world, lessons that they learned from their fathers and are passing on to their children. These are people who, even if they don't believe in man-made global climate change, are adapting to the consequences of a wild climate nonetheless. They're not doing it because of what they believe. They're doing it because of what they know, what's inscribed in their DNA. They're doing it because they have the tilth gene.

We need to listen to them, even if it's sometimes hard to hear them over the shrieks of protest from partisans on both sides of the climate debate who demand strict adherence to their almost religious orthodoxy.

{ TEN }

The Year the Creeks Stopped Freezing

IF YOU WERE TO FLIP BACK THROUGH THE LAST decades of acrimony and dispute to find the precise moment when the modern debate over climate and our role in it began in earnest, you would probably have to go back to June 23, 1988. That was the day that James Hansen, then a virtually unknown, buttoned-down middle-aged scientist heading NASA's Goddard Institute for Space Studies, leaned into the microphone at a table in a sweltering hearing room in the U.S. Capitol and told the Senate's Energy and Natural Resources Committee that the 98-degree temperature that was baking the sidewalks outside, a near record for the date, was, with a high degree of probability, only a pale shadow of what was to come.

"The earth is warmer in 1988 than at any time in the history of instrumental measurement," Hansen told the gathered senators, among them Al Gore, then a Tennessee Democrat who was carving out a niche as an environmental activist and was only just beginning to become a lightning rod for both sides in the harsh political and cultural battle that would materialize. Building on his agency's own research and on data collected over more than a decade by cloistered scientists across the planet, Hansen warned that there was enough compelling evidence that scientists were willing to state "with a high degree of confidence" that the buildup in greenhouse gases, primarily carbon dioxide from the burning of fossil fuels, was already affecting the temperatures on earth. What's more, there was evidence that the impact of the rising temperatures was "already large enough to begin to affect the probability of extreme events."

The next day, the *New York Times* ran the story with the headline "Global Warming Has Begun, Expert Tells Senate,"[1] that included this stark sentence: "[If] Dr. Hansen and other scientists are correct, then humans, by burning of fossil fuels and other activities, have altered the global climate in a manner that will affect life on earth for centuries to come."

It's doubtful that the fishermen who ply the waters off the coast of New Jersey out of the hard-luck port of Belford on the Raritan Bay took much notice of the story. None of them were the type to read the *New York Times*—they prefer the local newspaper, they'll tell you. And even if they had been inclined to read the *Times*, by the time the bundles of newspapers were dropped from the back of a truck at the local convenience stores, well before dawn, most of those guys were already on the water, forcing their small, rugged boats farther out to sea than they were ever meant to go, chasing the whiting they used to scoop up by the boatload just a few hundred yards offshore. Besides, there was nothing in that story, nothing in James Hansen's groundbreaking testimony, that they didn't already know.

Years before James Hansen and Al Gore issued their first warnings about how the changing climate was poised to set off a potentially devastating cascade of consequences that would likely accelerate as the decades passed, demanding both a reduction in the amount of carbon dioxide we consume and a concerted effort to adapt to the changing environment, the fishermen from Belford had seen it coming.

They didn't need the scientific treatises that Hansen relied on, or the scores of others that would come in the years to follow. By the early 1980s they could already see the changes. That was when the slow-moving, brackish creeks that carve through the marshlands at the edge of the bay stopped freezing over hard in winter and wouldn't again for decades, not until the much ballyhooed "Polar Vortex" briefly descended on the Northeast in the winter of 2013–2014. As Dave Diehl, a now partially retired fifty-three-year-old fisherman, told me years later, "I used to watch those creeks. The years they'd freeze over, that'd be a good year." Drawn to the cooler waters, the whiting, some as long as Louisville Sluggers, would swarm the waters of the Raritan Bay and nearby in the Atlantic. Those years, fisherman would head out to places like the old sludge dump, a mudhole so rich in nutrients that a whole food chain worth of species would congregate there just waiting to be caught. They'd haul them in in such numbers that their nets would groan under the weight. Back then, there were so many boats on the water that you could step from deck to deck all the way across the bay from Sandy Hook to Atlantic Highlands and never get your sneakers wet, Diehl told me.

And the years that the creeks didn't freeze? "You wouldn't catch shit," he said dryly.

By 1984, four years before Hansen's testimony before Congress, the creeks had stopped freezing. Scientists can and do debate the extent to which rising temperatures may have altered conditions in a place like Belford.

As far as the Belford fishermen are concerned, that debate ended a long time ago. "Nineteen eighty-four was the last good year we had," another old fisherman told me. It didn't take much. Since 1970, the temperature of the coastal water off the Eastern Seaboard has risen less than half a degree Fahrenheit, according to a 2012 draft report released by the National Climate Assessment and Development Advisory Committee, and while certainly not all of that could be laid at the feet of humans—some was part of natural variation—as far as the fishermen were concerned, it was enough. The whiting that once swarmed the coastal waters as regularly as the tides had chased the cooler waters far to the north and east, out where the big corporate trawlers and the fishermen from other regions prowled.

And even back then, the fishermen in Belford understood in

a way that scientists and economists and sociologists and world leaders are only now beginning to grasp that the issue of global climate change is not only maddeningly complex in its own right, but is tangled in an intricate seaman's knot of equally complex questions.

They had to worry about how warming land and sea temperatures were accelerating the rise of the seas, which have been creeping up at a rate of about an inch a decade, and could rise even faster as the ice sheets in Greenland and West Antarctica melt, or how the water in the Arctic would, in all likelihood, without the cold reflective cover of sea ice, soak up more and more heat, prodding those seas to rise faster and higher over time, as the land itself continued to sink. They had to concern themselves with the growing severity of the weather, how, scientists explained, warmer, wetter air that hovered over the ocean threatened to make deadly megastorms like Sandy—which claimed 159 lives in the United States alone, almost all of them within a couple of hundred miles of Belford—if not more frequent, at least more deadly. They might not have been able to do the calculations, but they knew as well as the researchers that waters were becoming more acidic as they soaked up the carbon we had been spewing, and like the researchers, they were beginning to wonder how that affected them. Soon enough, they would be hearing from fellow fishermen from the south in the warming waters of the Chesapeake Bay, who were beginning to see odd changes in the balance of sea life there—for example, the increased acidity seemed to be making oyster shells weaker, while the carbon that caused it seemed to be making the oyster's deadly enemy, the crab, grow to monstrous size.[2]

Because of where they stood, on a ragged patch of cattails 15 miles as the gull flies from Battery Park in Manhattan, at the heart of a massive and ever-sprawling urban and suburban juggernaut, a creeping mat of people and houses and cars and all the things that come with it, they could see how the warming temperatures were just part of the problem—a significant part, no doubt, maybe even the most significant in that it exacerbated and was exacerbated by all the others. They may not have been able to demonstrate it on a flowchart, but they could see how creeping suburban crawl, which by the mid-1980s was already surrounding them and by the late '80s was threatening to strangle them like a red tide on land, was taking its toll. More people didn't just mean more carbon; they meant more runoff trickling into their already embattled bay. It meant that the marshes and wetlands and barrier islands that had guarded the coast and soaked up the water and the fury of lesser storms were now being carved up, developed, and weakened just as forecasters were predicting stronger and even more lethal storms over the horizon.

And after countless twelve-hour days at sea, and sometimes weeks of twelve-hour days, battling larger, better-financed corporate fishermen for an ever-dwindling share of an ever-dwindling number of fish, watching as their own numbers dwindled, these guys understood in their aching bones just how hungry a beast the ocean could be. Out there at sea, they could peer off and see that the pressures that were being put on the water here were matched, even exceeded, by what was happening elsewhere in the world.

And they also figured out one other thing that the scientists and the economists and the sociologists were, as the years turned into decades, just dimly beginning to grasp: that in this tangled knot of conflicting imperatives and competing interests, in the Rubik's Cube cubed of coastal problems of which climate is one, any effort to tackle one of the problems, say overfishing, could make several of the other critical problems even more difficult.

Those who study such numbing complexities have a phrase they like to use to describe them. They call them "wicked problems."

The fishermen from Belford have another word for them. They call them their job.

That came into sharp focus for me on a frigid Friday afternoon in January 2013, when I stumbled into what remained of the old Belford Seafood Co-op, then the oldest operating seafood co-op in the region and one of the oldest in the country. I had been there before, and often, but that was a long time ago, more than thirty years ago when I was a young reporter working at my first real job, as a general assignment reporter for a scrappy harborfront weekly tabloid called the *Bayshore Independent.* In fact, I had from time to time been pressed into service to cover the war the fishermen were waging against a big developer who, seeing the potential in the grimy waterfront that others had missed, was trying to gobble up every acre of marshland and mud that he could to transform it into luxury housing. The fishermen back then had enlisted the aid of more than a few local coastal scientists and environmentalists, and ultimately convinced the Port Authority of New York and New Jersey, which had ultimate authority over the waterfront in that area, to spare them 10 acres so the co-op could remain afloat. I tried to be objective, but I can confess now that I was a little partial to the whole "Cannery Row" flavor of the place. And when the fishermen put on their only suits and rented a long white limousine to take them to the Port Authority's headquarters at the World Trade Center to sign the papers that seemed to protect their future, and then fired up cigars to celebrate, I stayed behind, but fired one up myself out of respect.

It had been decades since I had been there, and as I crossed the recently constructed bridge over the unfrozen creek and rounded the bend in the road that led to the co-op, everything was different. Gone were the old wood shanty and dock that used to sag under the weight of boxloads of fish. In their place was a long, low-slung green metal building that had obviously taken one hell of a beating, and fairly recently. And shouldering up on the place, within less than a hundred yards, where there used to be just salt grass and dunes, there were rows of $700,000 condos, going by some faux nautical name, leaving just a sliver of the old waterfront with its rotting ropes and rusting boats clinging to the side of the bay.

I parked my car and walked into the office, and before I could say anything, young Roy Diehl Jr., Dave's nephew and the son of the man who now serves as the co-op's president, a kid who

Roy Diehl with the Donna Lynn.

had just been born the last time I was there, greeted me with the words, "Looking for a job?"

At first, I was taken aback.

I'm fifty-four years old. The intemperance of the first half of my life and the general lack of moderation in all things that have defined the second half of my life so far have certainly carved their initials on my face. Certainly, I thought, he could tell just by looking at me that I wasn't some kid looking to learn the ropes of a new business. It wasn't until I sat down and talked with his old man, Roy Senior, for a while that I realized how perfectly reasonable young Roy's question was. The sad truth is that it's not at all uncommon these days for old guys like me to come knocking on the door looking for work. These are guys who spent their whole lives on the water, but who are now having a hard time trying to follow the dwindling number of fish out to the distant waters to which the increasingly warm waters close in were forcing them to flee. These are guys who couldn't navigate the crosscurrents of global climate change. Or global economics. Or government regulations, or catch limits policed by both the government and their own industry, regulations that, at least as far as the fishermen are concerned, were never really intended to keep up with the changes either. Instead they seem to have been designed to favor a handful of big corporate operations over the little guys, widening a gap between the haves and the have-nots that had always been there but now, driven by all these complex factors, was becoming deeper and wider and was swallowing them whole. In a lot of cases, these are guys who had their own boats and lost them, sometimes in seas farther out than those boats were ever meant to venture, sometimes to the banks.

If you want to fathom the depths of the wicked problem of which climate change is a serious part, you can start by gauging that mixture of sadness and respect in a thirty-year-old fisherman's eyes when he asks an old-timer, "Looking for a job?"

"That's my oldest son," Roy Senior tells me as the young man slips out of the office, toward the cracked concrete dock where the old man's 60-foot fishing boat, the *Donna Lynn*, rocks in a gentle swell. "He's been working with me since he was fourteen," Roy says, and he's one of a handful of young men, sons of fishermen, still doing what everybody around there knows is really a young man's job. The combined forces of economics and the environment have made it almost impossible for young people to get into the business, he says. He was, he admits, thinking about his son five years ago when he scraped together the down payment on the *Donna Lynn*, a boat he named after his wife. "I couldn't name her after my girlfriend," he told me. "That wouldn't have worked."

The boat had good bones but needed work, and every chance they got, after crawling back to port on his smaller old boat, Roy and his son would work well past sundown, laying down new decks, raising the gunwales, making sure that the boat would not only to be able go out farther but be fit enough to do it longer, for years into the future, so that one day it could carry Roy Junior, without his old man at the helm, out to sea. "The next generation will take over these boats," he says.

"After that?"

He just shrugs and tells me, "There's no way for them to take over. They can't go out on their own. That's the problem. You can't afford a boat, you can't afford the license and pay for it." Sure, you can spread those costs out over thirty years, he says. But the kids aren't blind. They've seen the changes that have taken place in the business and the bayshore over the last thirty years. "Who knows what the next thirty years will bring?" It's a $250,000 ante just to get into the game, and that's before you figure in the murderously high price of diesel, $1,500 a week for the average fisherman, he tells me. And for what?

Maybe it was easier, he says, when the water was still cool and the whiting still ran close to shore. Back then a guy like Roy Senior could readily make enough money to come up with a $30,000 down payment and buy a little house in what was then a forgotten corner of the coast, but that was before developers discovered the allure of bayside condos, and even now, after the real estate bubble burst, the local housing stock is out of reach for all but the most ambitious young fisherman. Roy Junior, just married and with a young daughter, he tells me, just dropped $200,000 on a fixer-upper and after working long, hard hours from before dawn to God knows when, he crawls home and works on that place.

Even after the creeks stopped freezing, there was a time when the fishermen out of Belford, or up from the nearby port of Point Pleasant, could still reap a bounty, heading out to an old sludge pit a few miles farther offshore than they used to go, where all manner of filthy castoffs from our modern society, pulverized into a thick mud, were dumped. The upside was that the sludge was full of nutrients that attracted creatures up and down the food chain. After a few hours on the water, the fishermen's nets would be bulging. The downside was that when the nets were empty, they'd have to tow them back to shore backward to clean out most of the debris and then power-wash out the rest. That debris, later analyzed, was found to contain a lot of "Kotex fibers and human hair," Jim Lovgren, a longtime Point Pleasant fisherman and a leading advocate for fishermen's interests, told me. So it was no surprise that by the early 1990s a coalition of environmentalists, with the support of a lot of queasy fishermen, successfully persuaded the government to shut it down.

But that meant that the whiting, the species that Belford had been built upon, no longer ventured even that close to shore, and fishermen like Roy Diehl had to follow them or chase species like skate, an ugly ray-like creature that some people say tastes a bit like shark but without the cachet, or fluke and summer flounder. But there, too, they ran into a headwind. Increasingly, environmentalists had been warning that the world's oceans were in danger of being overfished, that stocks of prime food species like the once-abundant cod that had drawn anglers to the waters off the coast of North America since the Medieval Warm Period that ended more than five hundred years ago were being depleted. So were the scallops and a host of other species, vacuumed up out of the waters of the Atlantic by large corporate fishing operations.

The small guys like Roy Diehl and Jim Lovgren disputed that. "You can never catch the last fish," Lovgren told me. If you did, you'd drive yourself out of business. Of course, the small fishermen operate on a different set of economic principles than the corporate operations do.

All the same, a tight net of regulations—some federal, some state, some established by the fisheries commissions themselves—was placed on the amount of each species that individual fishermen could catch, a portion going to the commercial fishermen, and another slice to the recreational fishing industry, which had grown exponentially in the preceding decades in tandem with coastal development, and those limits were rigorously enforced. It was not at all unusual for a government inspector to chug up to a small fisherman's boat to conduct an onboard audit of his catch. If the fishermen caught more than the daily limit of any species, he'd have to toss them overboard.

It puts a knot in your stomach, Diehl told me, to watch a guy who can't figure out how he's going to make his next boat payment dump $2,000 worth of perfectly good fish overboard.

What's more, because those quotas were doled out on the basis of past performance, fishermen were under immense pressure to keep up the pace. Take too few of any species, and there was a good chance that the next time the quotas were set, you'd be issued a smaller number.

It had become a kind of perverse incentive for fishermen to fish even harder, to risk going out on days when the barometer or the thermometer advised against it, to compete more fiercely, and always while running the risk of diminishing returns. As Timothy Fitzgerald, an analyst who studies fisheries for the Environmental Defense Fund, put it, the fishermen are "running a business and they are trying to maximize profits, and usually what fishermen do. This is not to their discredit. This is just the way that it works. They try and maximize their profits in the face of all of those restrictions . . . and . . . a lot of times it means that they get into this kind of cat-and-mouse game with the government, or with the management agency. So, if you say I can only fish ten days a month, I'm gonna fish with two boats instead of one. Or if you say I can only fish for something with one net, I'm going to make that net twice as big. Or if I can only be out one day a month, I'm gonna fish for twenty-four hours instead of twelve."

The result, he said, is that you end up at best treading water; you end up having the same or even greater fishing effort, even though the intent was to try and restrict it. That has economic consequences. It has environmental consequences. It has safety consequences for the fishermen. "You end up creating these derby fisheries or these races to fish where quality suffers, price to the fishermen suffers, because they all come back at the same time," he said.

And yet, despite all of that, there were still some decent years for the fishermen out of Belford, Diehl told me—not as good as the days before the creeks stopped freezing, but good enough to encourage them to hang on.

And then Sandy hit.

Like Belford itself, Joe Brennan has seen a few bad days. He's battled cancer and had hip replacement surgery, and he moves a lot slower than he did in the days when he chased whiting in his own boat, but he's still a Belford fisherman, even if he doesn't venture out onto the water much these days, instead spending his time as the office manager for the co-op. And in his thirty-six years at Belford, he's seen a few storms. He vividly remembers the 1992 nor'easter that pummeled the place for three days straight with winds of more than 100 miles an hour, and sent a foot and a half of water surging over the dock and into the cavernous green metal shed where the fish are kept on ice.

But nothing he had ever seen even began to compare with the raw murderous rage that was Sandy.

Scientists are still debating just how much impact the changing climate had on the fury of the storm that formed off the west coast of Africa on October 11, 2012.[3] To be sure, late-season hurricanes and tropical storms are nothing new along the East Coast. But there's ample reason for many of the experts to conclude that the unusually warm waters—the nation had been grappling with record-shattering heat all the previous summer—gave what would come to be called the megastorm additional fuel. And there is also strong evidence that changes in wind patterns high up in the atmosphere, changes that in all likelihood were driven by rising temperatures at the far northern latitudes, forced the massive storm toward the Eastern Seaboard, driving a storm surge before it that, when it crashed onto the shore along Raritan Bay on October 29 as part of a post-tropical storm, sent a wall of water and waves 8.9 feet above ground level—some of the deepest water recorded in the storm—hurtling past the tip of Sandy Hook, the long finger of sand that marks the dividing line between the ocean and the bay, taking dead aim at Belford and the bayshore. It was heading straight for developed coastline that had for most of the last hundred years been losing about one or two millimeters a year to rising sea levels, and now was losing ground at the rate of three to four millimeters a year, the fastest rate in 6,000 years, according to a paper published in 2013 in the *Journal of Quaternary Science* by Andrew Kemp and Benjamin Horton, a researcher at the Department of Earth and Environmental Science at the University of Pennsylvania.[4]

From the Carolinas to New England, the storm cut a swath of death and destruction. As a National Oceanic and Atmospheric Administration report on the storm released the following February put it, in the cold, stark bureaucratic language that such papers employ: "Preliminary U.S. damage estimates are near $50 billion, making Sandy the second-costliest cyclone to hit the United States since 1900. There were at least 147 direct deaths recorded across the Atlantic basin due to Sandy, with 72 of these fatalities occurring in the mid-Atlantic and northeastern United States," including at least one old man who drowned in his own bed. "This is the greatest number of U.S. direct fatalities related to a tropical cyclone outside of the southern states since Hurricane Agnes in 1972."

Entire communities, like New Dorp, a working-class neighborhood clinging to the edge of Staten Island, were virtually washed away. Out in Breezy Point, Queens, a community on the Rockaway Peninsula, in a truly apocalyptic image that was broadcast around the world, more than a hundred homes in a blue-collar neighborhood, places where the New York City cops and firefighters who were working around the clock helping others survive the storm had lived, burst into flames when the gas lines that served them ruptured, creating an island of fire surrounded by a raging sea. As many as 650,000 homes were damaged or destroyed by the storm, and for weeks—and in some cases months—millions would shudder in the cold and dark. Crews from all over the nation descended on the region, trying to patch together the aging power grid that served the powerhouse of the American economy, which had been twisted and snarled by the fury of the storm and tossed into the surf like the iconic roller coaster at Seaside Heights in New Jersey.

In Belford, the fishermen rushed to salvage what they could before that wall of water rolled over them, first tying down their boats. That's one of the things a fisherman knows how to do, and while uncounted pleasure craft owned by weekend anglers were ripped from their moorings and some of them dragged a mile or more inland, the fishing fleet at Belford didn't lose a single boat.

But it didn't escape damage. As the storm surge crashed ashore, more than seven feet high by the time it hit Belford, it peeled away the metal skin from the quayside wall of the co-op's headquarters, and as it receded it knocked out the sidewall and gutted a little restaurant, its current so strong that it carried a commercial stove, a giant refrigerator, and every utensil in the place back out to sea with it.

And even when the skies cleared and the surge retreated, the storm still wasn't through with them. The physical damage to the structure was bad enough, Roy Senior told me, but even though the co-op didn't have flood insurance—it couldn't afford it even in the best of times—it did have wind insurance, and that covered some of the damage. And of the billions spent during the first halting steps to restore the devastated region after the storm, the Belford co-op managed to snag $11,000, about the cost of a week's worth of diesel for all of the co-op's five or six remaining full-time fishermen.

The problem was, for the time being they had no need for the fuel. The storm had fried the co-op's electrical system. "No electrical, no ice," Roy Senior explained simply. "No ice, no fish."

It would end up costing each of the fishermen between $40,000 and $50,000 to rewire the place, to get the freezers going so they could get back onto the water, back into that maelstrom of a thousand competing problems of which climate was only one, in order to make a living. For two long months, they were grounded.

And in those two months, the newspapers—the *New York Times*, the *Newark Star-Ledger*, the *Asbury Park Press*, and others—would run story after story detailing not just the horrors Sandy had left in its wake but also the increasingly strident debate about the role that climate change may have played in it. There were stories pointing to ever-improving scientific models that seemed to indicate that regardless of how much—or how little—climate change may have contributed to Sandy's rage, there were strong indications that as long as the oceans continued to rise and the atmosphere continued to warm, savage storms like

Sandy were likely to become more common, and things were likely to get a lot tougher for guys like Joe Brennan and Jim Lovgren and Dave Diehl and Roy and his son.

The writers always saved a little space for those who claimed the models were bunk—in the interest of balance, you understand. And then they'd go on to write about how it was time for elected officials to talk seriously about planning for the next time, maybe redesigning buildings and infrastructure, getting them out of harm's way, so that next time the damage might be less severe.

And there were stories about others who refused to even entertain the notion that climate change was real or who talked of defiantly rebuilding everything right where it had been when the storm hit. There were stories like the one about the little oceanfront town of Avon-by-the-Sea, a few miles south of Belford, that planned not only to build its boardwalk right on top of the pilings left behind when the old one washed away, but to build it out of ipe, a wood harvested from the Amazon rain forest, the world's largest carbon sink. Why? Because they could. And that's what they had rebuilt with the previous time.

The guys in Belford were busy repairing their damaged co-op, rewiring the place, laying the groundwork to build a new office, this one on the second floor, in the hope that retreating to higher ground would make them better able to defend themselves the next time a massive storm surge barreled over their dock. Maybe it wouldn't happen next year. Maybe it wouldn't happen for five or ten years. But it would happen, they told me, as sure as the sunset, and they were going to be as ready as they could be.

Of course, that didn't leave them a lot of spare time to sit around reading newspapers.

Besides, nothing in those newspapers would have told those guys anything they didn't already know.

They'd known it for years. In a way, they'd known it ever since the creeks stopped freezing over.

A road sign in drought-stricken West Texas.

{ ELEVEN }

"It's What I Do"

It wasn't a torrential rain, nothing like the near-record-breaking deluges of the spring of 2013 that would wreak havoc on the Midwest a few weeks later, flooding farmland, sending the Mississippi and Illinois Rivers surging over their banks in many places, and claiming five lives from Michigan to Missouri in the process. But this gentle rain was enough, even in late March, to keep Ethan Cox off his tractor and to hold him off on planting for a few more weeks.

And as he sat there on his overstuffed couch listening to the rain pelt out a dismal tattoo on the metal roof of his workshop, Ethan had time to think.

Maybe it wasn't global warming, he told me, these wild swings in the weather that were turning his neighbors' fields—fields that just a year earlier had been baked to the consistency and color of terra-cotta—into lakes, washing away the precious topsoil and robbing the farmers of the critical weeks they needed to get their crops in. He still wasn't sure he believed in it. Maybe it was.

To him, it didn't matter. What mattered to him was that he was doing all he could to protect his land, his farm, his daughters' legacy, from the ravages of what even he acknowledged were increasingly volatile storms and increasingly savage droughts that seemed to him, and to a good number of the scientists who are studying it, to be coming more frequently.

A few days earlier he had taken a ride and seen the deep ravines cut in his neighbors' fields, his neighbors with the tire gene rather than the tilth gene, and it convinced him that he had done the right thing. Not just in the sense that it was the correct thing to do for himself, but also in the sense that the choices he had made were the right ones ethically as well.

"It's a sin," he had told his daughter Maria, "letting it erode like that." It wasn't a figure of speech. He literally meant it in the theological sense, though he would never describe it that way himself. You'd have to poke him and prod him a bit to get him to explain, but what he meant was that hubris and carelessness were taking a toll on the land, and as a result, people far beyond the farms themselves would be affected, through higher food costs, through the thousand ways we all bear the costs of disasters that are made worse than they need to be.

Eight hundred miles to the southwest, in David Ford's parched fields outside of Dumas, he would tell you the same thing if you could find the right way to frame the question. The steps he had taken—adopting minimum tillage, strip tillage, reducing his consumption of oil while at the same time increasing his yield—were not just practical in an agricultural and economic sense, they were ethically correct as well.

Would those steps turn back the clock? Could they alone replenish the overburdened aquifer underneath Ford's land or claw back the carbon that has already locked us into a roughly one-degree increase in the global temperature that has contributed, as John Nielsen-Gammon told me, to making the drought in Texas, the weather extremes in Illinois, and the fury of storms like Sandy off the East Coast even worse?

No, of course not. But that's not to say that they're useless. For Cox and Ford, they're buying time for their few thousand acres, a few more years, maybe, time enough for their children—Ford's sons, Cox's daughters—and for the young ranch hands at the 6666 Ranch to hold on in the hope that someday we'll be able to get beyond the harsh partisan divisions in this country and take real steps to slow the march of rising temperatures.

It's easy to think that they're deluding themselves, that whatever impact they might have is insignificant. But even in the most practical sense, that's perhaps too cynical a view. The truth is, the vast majority of farmland in the United States—87 percent of the roughly 2 million farms in the nation, according to the 2007 Census of Agriculture—is family farms.[1] Only about 4 percent are so-called "corporate" farms, and even that number includes partnerships and tightly held corporations that are often held by families. And while only a fraction of those 2.2 million farms—187,816 of them—produce 63 percent of the agricultural products we consume, that number includes a significant number of large family-held farms like Ford's 7,000-acre operation or Cox's 3,000-acre farm. For that reason, says Robert Hoppe, an agricultural economist for the U.S. Department of Agriculture and one of the authors of the agency's 2011 report *The Changing Organization of U.S. Farming*,[2] the kind of experimentation that Ford and Cox and thousands of other farmers and ranchers like them across the country are doing has the potential of pioneering at least a piecemeal and temporary response to some of the risks posed by climate change.

"These guys are central," he told me bluntly.

As the 2011 report put it, in the face of dwindling water resources in some regions and potentially devastating deluges in others, "practices such as the use of genetically engineered seeds and no-till have dampened increases in machinery, fuel, and pesticide use. Likely aided by the increased use of risk management tools such as contracts and crop insurance, U.S. agricultural productivity has increased by nearly 50 percent since 1982."

As the challenges become more severe, "future innovations will be necessary to maintain, or boost, current productivity gains in order to meet the growing global demands that will be placed upon U.S. agriculture," the report concluded. And by dint of their sheer numbers, those innovations are most likely going to come from the farmers in the field rather than directly from some corporate laboratory, Hoppe told me.

Of course, Cox and Ford and Joe Leathers at the 6666 Ranch

don't speak in those terms. They are not trying to save the world. They're trying to save their own farms and ranches. As Cox told me, it's all he has to give to his children. It's all he knows. It's all he's ever done, and everything he owns in the world is tied up in the place. His net worth may be $3 million or $4 million, he says, but all of that is in land and equipment, things that have value only as long as they're useful, and as long as they're useful, he'd rather be the one using them, him or his daughter.

Sure, he admits, he may be a dinosaur in some respects, a throwback to a distant time. And there may come a time when guys like him can no longer make it. "I always had the fear that I wouldn't be able to farm," he told me. That would be a tragedy for her, his daughter Maria said, but it wouldn't be the end of the world. "This is who I am," she said. "But I could do something else." It would be a catastrophe for her father, she said.

I found myself thinking back to what Mathew Barrett Gross, the coauthor of *The Last Myth*, had told me on that Sunday morning. Every day is the end of the world for somebody, he said. And it always has been. Ever since we first climbed down from the trees, we've been one step ahead of disaster, famine, flood, wars. When we all lived in thatched huts, a single lightning strike or a motivated guy with a torch and a bad attitude could usher in what certainly would have seemed to those people directly in its path like the end of days. And yet we survived; in fact, we thrived.

And maybe we did because we believed we would.

Not long after my chat with Gross, I found myself in a windswept corner of West Texas on a hundred-degree day, making my way from Ford's place to the 6666 Ranch, 60 miles to the south.

The landscape was parched, and behind fences that had once penned in vast herds of cattle, there was nothing, nothing except piles of mesquite and brush stacked as tall as a man. I had stopped in at Sarge's, one of those desolate roadside places in the middle of nowhere that sell whatever is needful, and ended up chatting with an old man inside. I asked him about the piles of brush. Some of it had been burned. Some of it was waiting to be burned. He explained to me that the ranchers were fighting a never-ending battle to keep their pastures clear of brush in the expectation that sooner or later they'd be able to bring their herds back. Those fires were a testament to the hope that by simply getting up every morning and doing what they knew how to do, eventually they'd be able to make even a small difference. And if not, at least they would be able to say that they'd done something.

I saw it at the 6666 Ranch when I caught up with young Steve Briggs, a ranch hand who was making his last-minute preparations before leaving on his two-week deployment to Montana to look after what remains of the ranch's once-massive herd. Though it had been nearly two years since the herd was moved away, Briggs had been working steadily, both in Guthrie and, when needed, in Montana, because that's what he does, and because he absolutely expects that someday the herd will return. And in the meantime, the land still needs to be tended.

The small steps that the farmers and ranchers are taking are, of course, not anywhere near enough to avert what most of the scientists contend are the worst potential impacts of a changing climate. But what they may be are examples of what religious historian Richard Landes was talking about when he told me that the apocalyptic doomsayers who dominate our public debates always miss the one key element that distinguishes us Americans and has, so far, always led us away from the brink of disaster: the fact that we are ultimately a hopeful people who believe that we have the ability to change our fate.

To be sure, it'll take more than a few farmers and ranchers and fishermen to do it. It will take leadership from government, and from industry, to tackle the kinds of large-scale efforts that researchers say will be necessary to fortify our vulnerable places against the extremes of weather linked to climate change and to reduce the amount of greenhouse gases we release. And that leadership has been in short supply.

Or maybe it only seems that way because we're looking in the wrong place for leadership. Maybe Washington and Austin and

Springfield and Trenton aren't where real leadership will emerge. Maybe we should be looking in White Hall, Illinois, and Dumas, Texas, and Belford, New Jersey.

At the moment, those ranchers in Texas, the farmers in the Midwest, the fishermen off the coast of New Jersey are the voices we seldom hear. They're drowned out by the screech of harsh partisans on either side. And in that din, far too many of us just cling ever more desperately to whatever fragment of information gives us solace.

And yet despite all of that, there is evidence that a growing number of Americans are starting to feel that they've been betrayed by the extremists on both sides of the issue. These are people who often are not ideologically, politically, or in some cases even religiously inclined to accept the political baggage that goes along with discussions of global climate change, but who, often heroically, are dealing with those consequences nonetheless and are beginning to realize that they need to demand leadership from both state and federal governments and accountability from the industries that are exacerbating the crisis. And though they are still in the minority, these are people who are in a way pioneering a different paradigm for discussing the issue. They see the world changing, and are changing themselves, in some cases by adapting to it, in some cases mitigating their contributions to it, always driven by practical concerns. In so doing, they may be showing our policymakers and the pundits that there is another way to discuss the issue of climate. If only our leaders are smart enough to recognize it. Or if we're strong enough to force them to recognize it.

"Congressmen don't lead, they follow," Bob Inglis, the former South Carolina congressman turned conservative climate crusader, told me when we chatted. Right now, they're following the money, the media, the harshest voices on either side of the cultural and political fractures that snake through this country. But maybe that can change.

It's worth noting that Ethan Cox's congressman, Republican Rodney Davis, is on record as stating that climate change, if it ever was real, stopped in the 1980s. The congressman for the 13th Congressional District in Texas, which includes both Guthrie and Dumas, is Mac Thornberry, who famously said that prayer was perhaps a better response to the devastating drought in his district than reducing global emissions of greenhouse gases.

Maybe Davis and Thornberry and other members of Congress who have taken a similarly see-no-evil approach to the challenges their constituents face should listen to what men like Cox and Leathers and Ford have to say. Maybe they should listen to what Hayhoe and Tanner have to say. They should certainly realize that the born-again scientist and the Montana outdoorsman are speaking to—and increasingly speaking for—a growing number of their constituents. These elected officials needn't change their beliefs—not that changing beliefs is unheard of among congressional representatives. After all, Cox and Leathers and Ford themselves are far from convinced that climate change is the cause of all of their hardships. They don't need and may not even want their elected officials to profess a belief in anthropogenic climate change. What they do want is for their elected officials to believe in them. And in exchange, what these guys offer is an example—an example of a defining American trait, a willingness to overlook the fractures in the political and cultural landscape, to listen to what their land is telling them, and to respond to the challenges with the hope, however far-fetched it might seem to those of us who are detached from the land, that native ingenuity and hard work can change things. Even if only a little.

We've certainly done some serious damage already. In May 2013, the researchers at the Scripps Institution of Oceanography announced that their observatory at Mauna Loa in Hawaii had reported that for the first time in history, we had crossed a daily threshold. The daily readings showed that we had reached a carbon level of 400 parts per million and that we had reached it far faster than most scientists had expected. "I wish it weren't true," project director Ralph Keeling said in a statement, "but it looks like the world is going to blow through the 400-ppm level without losing a beat. At this pace we'll hit 450 ppm within a few decades," the number at which Mann and other scientists warn that potentially devastating impacts of the changing climate will be locked in for a very long time to come.

It may well be that there never will be a magic bullet, no single solution or package of solutions that will successfully fend off the worst effects of rising seas and harsh droughts. Sure, we can slow the process if we have the will to reduce our consumption of coal and oil, wean ourselves off natural gas over the next few decades, limit the environmental impact of biofuels, and expand our array of renewable power sources. Maybe we can cut our consumption and buttress our infrastructure a little bit at a time. But it may well be that we've already loaded the skies with enough poison that the best we can hope for is to manage the problem, at least for the foreseeable future, and maybe find a way to tough it out.

And if that's the case, maybe we all need to learn a few things from guys like Ethan Cox and Roy Diehl and David Ford and Joe Leathers, guys who have been managing their little corner of the problem for years now.

{ TWELVE }

Penguins Tumbling Off an Ice Sheet

THE THING THAT STRUCK ME MOST ABOUT RICHard Alley, the noted climate scientist, when I visited him at his Penn State office was his utterly unflagging sense of optimism.

No. That's not quite true.

The thing that struck me most was the absolutely childlike joy he was experiencing when I walked into his office and found him cackling with abandon at a video he had taken a few weeks earlier at the bottom of the earth. It showed one particularly determined penguin, in a flock of thousands, that could not, for the life of it, find a way to keep its footing on a slick ice sheet. I have to admit, I found it a little jarring to hear this respected scientist guffawing wildly as he watched the young penguin struggle to remain upright, get pushed off, and in what clearly seemed to my human eyes to be a heroic effort to mask its injured pride, straighten its feathers and try again. And again. And again.

After all, if anybody in the world has reason to be dour and pessimistic, it's Richard Alley.[1] For more than a decade he's been among the world's most cited researchers into the challenges of global climate change. He was one of the principal authors of the 2007 IPCC report, after all, and most of his waking hours are spent either studying the data, lecturing his students about it, or trying to lay out the problem to lawmakers. As an author and host of the PBS series *Earth: The Operators' Manual,* he has been

Richard Alley on the Matanuska Glacier in Alaska.
Photo by Todd Johnson.

among the leading voices warning about the dangers of melting ice caps and sea level rise.

And yet, Alley, who was once described by veteran *New York Times* environmental journalist Andrew Revkin as "a cross between Woody Allen and Carl Sagan," is an astoundingly optimistic man, even if he does take an unusually enthusiastic interest in the antics of Antarctic waterfowl.

Just looking at the raw numbers, it's hard to imagine where Alley's optimism might come from. After all, the first ten years of the century were, according to a 2013 report by the World Meteorological Organization,[2] the warmest decade in the 160 years that such measurements have been recorded. Worldwide, 2010 was the hottest year on record, and two years later, in 2012, the same year that Ethan Cox battled the drought in Illinois, the United States topped its personal best—or worst—reaching an average temperature of 55.3 degrees Fahrenheit, a full degree warmer than the previous hottest year on record, 1998. That year, records fell like cornstalks in the heat. A total of 34,008 daily records for high temperature were set at local weather stations across the nation, with only 6,664 records for lows, according to federal temperature records.

That same decade saw staggeringly destructive weather extremes around the globe and in the United States: the record heat wave of 2003 that was linked to the deaths of as many as 35,000 people in Europe, 14,000 of them, many elderly, in France alone, and the 2010 floods in Pakistan—the worst deluge to hit that region since 1929—that claimed 1,600 lives, according to a United Nations report, and devastated that nation's food supply. And of course, there was Hurricane Katrina, with winds of 125 miles per hour, which hit the Gulf Coast on August 25, 2005. That storm killed 1,577 people in Louisiana, 238 in Mississippi, and 18 more in Florida, Georgia, and Alabama, devastating the city of New Orleans and making Americans question in a way many of them never had whether their government, a government that has, if anything, become even more paralyzed by polarization in the years since, was up to the task of protecting them against the worst ravages of extreme weather.

In each of those cases, and in the savage storms and droughts that followed in the next several years, a number of factors at play made the consequences more severe. In our shortsighted way, we had allowed our infrastructure to deteriorate. Even as the temperature was rising decade by decade, there was an almost willful failure to gird vulnerable people and places against the increasing likelihood of extreme events, and an embrace of almost unbridled development that encouraged building in risky areas—post-and-beam McMansions in Todd Tanner's fire-prone western forests, condos on the waterfront a hundred yards away from Roy Diehl's office at the Belford Seafood Co-op on New Jersey's vulnerable Raritan Bay—a painful reluctance on the part of bureaucrats and farmers with the "tractor gene" to listen to their land the way Cox had.

The Government Accountability Office reported in 2013[3] that the federal government was ill-prepared to bear the financial burden of extreme weather events linked, even peripherally, to global climate change—asserting that "while it maintains vast areas of publicly owned land, is responsible to millions of flood and crop insurance policyholders and spends billions of dollars in natural disaster response and recovery aid every year, the federal government still lacks any coordinated system for dealing with the financial costs of climate change and global warming." In fact, House Republicans, still officially hostile to anything related to climate, voted to strip $650 million from NASA's Earth Sciences Division, which had been studying climate change since before James Hansen warned Congress in 1988 that the climate was indeed already changing around us.

And yet, as Michael Mann, Alley's Penn State colleague, put it in his 2012 book, *The Hockey Stick and the Climate Wars: Dispatches from the Front Lines*, the average American continues to consume vast quantities of energy from fossil fuels. Despite the sluggish growth that still hobbles the economy well into the second decade of the century, despite significant advancements in energy efficiency with the promise of more down the road, despite the promise of renewables—Texas alone had deployed 12,214 megawatts of wind power by 2013, triple what it had a decade earlier, and Pennsylvania, in the throes of its own fossil fuel boom with the Marcellus Shale, has deployed 795 megawatts of wind power, with a target of 900 in the next several years—Americans on average each continued to emit roughly 20 tons—the weight of two very large adult male African elephants—of carbon per year.

And while we may be among the planet's most voracious consumers of energy, battling China for the title of the world's largest power glutton—we are certainly not the only ones. As Mann put it, "Globally, human beings emit the equivalent of nearly 400 million of those elephants—8.5 billion tons" annually, a number that is likely to increase as the world's population continues its march from 7 billion people, 1.5 billion of whom currently have no reliable source of energy, to perhaps as many as 9 billion by the middle of this century.

Even if we were to slash our consumption by 50 million elephants by the middle of the century and to near zero by the end, Mann contends, there is nevertheless enough trash in the air that we would likely still see the temperature rise by another 1 degree Fahrenheit and see sea levels a foot higher at the Belford Seafood Co-op and at my mother-in-law's house on Florida's Indian River Lagoon over the next century.

And even as we face that risk, we continue to pump vast quantities of fossil fuels out of the ground. Although coal has lost a significant part of its market share in the United States as a fuel for electrical generation in the past decade or so, dropping from about 50 percent of the market between 2001 and 2008 to a low of about 35 percent in 2012, it remains robust.

And then, of course, there's natural gas and vast caches of oil, oil from shale and tar sands, that have been unleashed by new technologies, hydraulic fracturing among them. By 2010, the thick, costly-to-process tar sands of the Athabasca region in Canada, a hundred thousand square miles of forest north of Alberta, were thought to contain the equivalent of 1.7 trillion barrels of oil, or about a fifty-year supply at the current rate of consumption, said Michael Levi, an energy analyst at the Council on Foreign Relations and the author of the 2013 book *The Power Surge*.[4]

It's small wonder, then, that in early 2011 James Hansen—the same NASA scientist who had first alerted Congress to the risks of climate change—warned that if all the deposits in Alberta were burned it would be, in his words, "game over" for the environment.

That warning inspired a movement, not just to block the development of the tar sands, but also to at least impede the development of other fossil fuel plays, from Alaska to the coast of Virginia to the Utica Shale in Ohio. As activist Bill McKibben wrote in the July 19, 2012, issue of *Rolling Stone*,[5] "Scientists estimate that humans can pour roughly 565 more gigatons of carbon dioxide into the atmosphere by midcentury and still have some reasonable hope of staying below two degrees [Centigrade]." If all the proven reserves were burned, McKibben wrote, citing the work of the London-based environmental and financial analysts at the Carbon Tracker Initiative, we'd send another 2,795 gigatons of carbon into the atmosphere.

Of course, that's assuming we burned it all.

The way Levi sees it, it's possible but not particularly likely. "What matters is not how much oil is in the ground; it is how much will be burned," Levi told me, effectively quoting his book. It would, he argued, take three thousand years to burn the bitumen in the tar sands alone. What's more, he said, even if we doubled the amount of oil we produced, all of that additional oil flooding onto the world market would likely have an effect on prices, in all probability driving them down to the point where either foreign producers like OPEC slashed their production rates or the bitumen from Alberta, which is costly to produce, was no longer worth the investment in money or time.

Levi's argument was bolstered in mid-2013 by a study from a team of researchers at Stanford University led by Adam Brandt, "Peak Oil Demand: The Role of Fuel Efficiency and Alternative

Fuels in a Global Oil Production Decline,"[6] which opined that while we may no longer have to worry about hitting "peak oil," we may very well reach a point of "peak oil demand" by the middle of the century.

But equally important is the role likely to be played by increased alternative fuels—a category that includes low- and no-carbon fuel sources, but also, perhaps perversely, carbon-spewing fossil fuels like tar sands and shale oil. As Brandt put it in an e-mail exchange with Andrew Revkin of the *New York Times Dot Earth* blog, "The particular path taken away from conventional oil will strongly affect the climate consequences of our future energy systems. A shift to wind and solar-powered [electric vehicles] will obviously have climate benefits. A shift to coal-based synthetic fuels may mitigate the need to burn conventional oil but will exacerbate climate change. The complex social and ecological challenges of biofuels are well known."[7]

But despite the thorniness of the questions raised by the options for alternatives, there is still, in both Levi's and Brandt's analyses, cause for some hope, hope that was buoyed in 2013 by a series of reports that suggested that the worst-case scenarios—those calling for a six-foot increase in sea level, for example—may be less likely, at least over the next century.

That's not to say that Greenland's ice sheets are going to stop melting into the sea, or that Antarctica is going to stop calving off Manhattan-size icebergs, but it does mean that crashing through the psychological barrier of 400 parts per million of carbon in the atmosphere is not the end of the world. It means that we may have more time, if only slightly, to mitigate the rising temperatures, at least partially, and adapt to the risks we've already created.

"There is," as Mann put it to me that day in his office, "no magic number."

He continued, "I wrote a piece [for] the *Proceedings of the Academy of Natural Sciences* a few years ago where I argued that it's reasonable for a number of reasons to call 2 degrees Centigrade warming relative to preindustrial [levels] a threshold beyond which impacts become really severe. In reality, I think it's a continuum . . . it's a slope, and how far down do you want to go?

"It doesn't mean that if you miss the exit ramp, you don't get off on the next one."

To be sure, we are accelerating toward the point where we've loaded the atmosphere with enough carbon to affect the climate for a very long time to come. After rising relatively slowly from 1850 to the 1970s, carbon levels increased dramatically from 361.5 parts per million in 1991 to 400 in 2013. And increasingly, scientists like Mann are eyeing a number around 450 parts per million as the point beyond which "it's pretty clear that the damages outweigh the costs of mitigation," he said. "A reasonable person can conclude that we probably don't want to go there, beyond 2 degrees C."

And an eminently reasonable person a few hundred yards away from Mann's office remains blissfully optimistic that we won't.

It's not that Richard Alley has any illusions about how rapacious we humans can be. "You're sitting here at Penn State, and Penn State was built by the iron masters. . . . We're a suburb of their iron furnace, and that iron furnace took more than a half mile of trees every year just for charcoal. We clear-cut Pennsylvania. Penn's Woods had no trees, essentially.

"I visited the whaling museum in New Bedford. We cleared the Atlantic of every whale we could catch so we could burn them at night and have lamps. I've just been down to the whaling stations in South Georgia and saw what happened after that. We were just cleaning the world out of whales. And now we're cleaning the ground out of gas and oil and coal."

And yet, that cold-eyed realism has done nothing to douse Alley's almost irrepressible optimism about the future. "If you want a really wonderful, powerful vision, it is that we now know how to do this, we know how to build an energy system that will power everybody on the planet forever, and forever is a million years. We know it's technically and engineering-ly feasible at this point, and we know it's within the economic realm."

As he said to me that day while he was cackling at the penguins, "There's an old joke—I think you will hear this among the Alarmed. A guy falls off the roof and he's falling past the windows and he keeps saying, 'So far, so good.'

"I think that's the wrong analogy," he said.

It's certainly not as dramatic a metaphor, but the way Alley sees it, the issue is more like saving for retirement. Every day you delay is costing you, but whenever you decide to start, it helps.

Certainly we can't just snap our fingers and expect it to happen, he said. It will require compromise, half steps, occasionally a step or two backward, the kind of deals that fatalistic absolutists on either side of the political debate over climate can't tolerate. But Alley has an unflagging belief that we are far more likely than not to blunder toward a solution, because, he argues, that is precisely what we have always done as Americans. "You compare the history of the ancestral Pueblo [the people of the American Southwest] and you compare it to the history of the population in the valleys there . . . and when it rains there's more people, and when the drought hits there's fewer people, and when the really bad drought hit[s], they're gone. Packed up. And you compare that with what happened in the Dust Bowl."

Under assault from the combined impacts of bad farming practices and a significant drought, the land simply dried up and blew away, forcing thousands of sod-busting farmers from the Midwest to West Texas to abandon their homes and farms. "A lot of Oklahomans packed up and left," Alley said. "But if you look at the population of Oklahoma, it didn't drop much. Most of them hung on. And then you look at the next drought that hit and the population doesn't drop at all." The farmers adapted. They mitigated. "They figured it out."

That is, Alley told me that day, the grand, sweeping message of American history. "With absolute certainty we know how to get screwed by the climate and we know how to screw ourselves. And we know all of these histories where people failed."

To be sure, Mann and Alley both believe the clock is ticking, and they are certain that we have already churned out enough greenhouse gases to lock in formidable challenges well into the future, for hundreds of years, perhaps thousands. "But it's not that if we wait ten years then everything is doomed, and we're all going to die," Alley said. "I think we're still weeds."

And like weeds, we have an astounding ability to adapt, he argued, pointing to the history of whale oil as a case in point. For the first five decades of the nineteenth century, the world pursued a devastating campaign to squeeze every drop of oil it could out of the carcasses of the world's whales, driving some species to the brink of extinction. By the mid-1840s, the United States alone had 735 whaling vessels in its national fleet and was producing up to 14 million gallons of whale oil annually. But as the number of whales in the sea dwindled—as we hit "peak whale oil," so to speak, and began the quick, steep descent into scarcity—the price tripled.

It took a few years for the nation and the world to grasp the consequences of the whale oil crisis, Alley said, and during that period there were financial hardships, but eventually, oil—the stuff that came gushing out of Colonel Edwin Drake's first well in Titusville, Pennsylvania—and coal supplanted whale oil. "The free market ultimately did work," he said, "but there were lots of . . . bumps . . . along the way."

It is true that the recent technological advances that have unlocked vast stores of oil and natural gas that were previously out of reach have expanded the amount of oil available to us to burn, and it also true that there remains a serious challenge—socially, economically, environmentally—in doing so. A 2013 study by the White House[8] concluded that the burning of fossil fuels adds billions of dollars in costs to the consumer, costs that are hidden when we fill up our gas tanks or crank up our air conditioners. Those hidden costs include disruptions to food supplies brought about by savage weather made all the more extreme by the changing climate, the kinds of challenges that Ethan Cox and Joe Leathers and David Ford are grappling with. There is the kind of destruction wrought by increasingly violent weather like the superstorm that ravaged the coast of New Jersey and grounded the fishermen from the Belford Seafood Co-op for months in the fall of 2012, and the billions in losses associated with that. There is the cost of responding to the threat of rising seas that the Army Corps of Engineers is calculating in Florida and that Klaus Keller is trying to calculate everywhere.

At the moment, Alley argued, even with a hopelessly gridlocked Congress, and only the first tentative steps toward leadership on the climate issue from the administration, we are making progress despite ourselves. Coal consumption has declined, renewables have expanded dramatically, and energy efficiency has slowed our seemingly unstoppable rush to burn more oil and fossil fuels.

Certainly, he said, alternative technologies that now account for only a fraction of the energy we consume have a long way to go. We can't just hope that wind turbines will magically appear—and, of course, only on those mountaintops where they don't impede the view. We can't just expect that solar arrays will appear without the carbon it takes to mine the raw materials for them, to smelt them, to manufacture them, to transport them and erect them in place, and to supplement them when the wind doesn't blow and the sun doesn't shine—at least not until we develop a cost-effective and reliable means of storing the energy they produce.

And so, both Mann and Alley argue, for the short run we're going to have to compromise. Those who insist that we do nothing are going to have to come to terms with the fact that inaction is no longer an option, that continuing down the path we've been pursuing will have profound costs, costs that are already becoming evident.

And those who oppose the development of marginally cleaner technologies like natural gas, or significantly cleaner but still risky options like nuclear energy, embraced by James Hansen, will also have to deviate—if only slightly—from their sacred texts, Alley said.

To be sure, there are risks, he said. "I mean, gas is . . . particularly problematic in one way because it actually does represent an increase in the amount of fossil fuel that's available and if you simply burn everything else and add this on top, we get more climate change," he said. "On the other hand, it's quite clear that properly done it has less of an impact than the things that it's replacing. . . . You can see this as a bridge to a truly sustainable future."

As Mann put it, "Nuclear energy . . . poses serious risks. But they are very different types of risks and the way that we can mitigate and sort of insulate ourselves from risk is very different there, and the time frame in which the risk plays out is very different. That's true even within fossil fuel. Natural gas? I was recently . . . quoted . . . saying that it was sort of cautiously optimistic news when it was announced . . . that we had seen a decrease in our carbon. People can debate what really underlies that, but most people accept that natural gas, a transition to natural gas from coal, probably played some role in that."

The risks are real. Some of them may be manageable. Some of them may turn out to be more trouble than they're worth. And both men argue that all of the potential solutions raise a host of complicated questions.

At the moment, neither is prepared to fully endorse any single technology to address the rising levels of greenhouse gases in the atmosphere.

What they are prepared to endorse is more—and better—communication on the issue.

What Alley endorses—as a scientist—is a fact-based consideration of all the risks and all the benefits, and what he also endorses—as a hymn-singing churchgoer—is keeping that debate as respectful as possible.

The idea that in this polarized country we could have such a respectful discussion may seem naive, but that is precisely what people like Bob Inglis and Rob Sisson and Katharine Hayhoe and Todd Tanner are trying to tap into. In essence, they are trying to harness that most renewable of American resources, a sense of hope, a sense that has historically motivated us, a belief that we are a people who, in a quote widely misattributed to Winston

Churchill, can be counted upon to do the right thing after we "have exhausted every other possibility."

It dawned on me as I left Alley's office that day that perhaps that was really what he found so funny about the antics of the penguins he had filmed on the ice sheet in Antarctica. The fact that one penguin, despite its repeated failures to find the one place where it could stand without being forced off the edge, kept getting up, grooming its feathers, and trying again may tell us as much about our own way forward as all of Alley's work in the Antarctic tells us about our certain past and our potentially devastating future.

Acknowledgments

THE POWER HAD GONE OUT, EVEN BEFORE THE storm that had slammed into the Eastern Seaboard with such ferocity washed up on the eastern face of the Pocono plateau, and for a few hours my kids and I huddled around the fireplace. Just waiting.

And when the storm finally did hit, I left my kids downstairs and stepped out onto the front deck. It was surreal. Winds in excess of 80 miles an hour were howling, but because of a trick of geology and an accident of design—my home is built halfway down a gorge in the eastern Poconos—the winds never touched me. I could stand on my deck and hear them, a hundred feet over my head, roaring like a coal train as they hit the ridge across Bushkill Creek and were deflected ever so slightly over my house.

I remember thinking that it was as if somebody had painted an *X* in lamb's blood on my doorposts.

Neighbors nearby were not so lucky. Ancient oaks and hemlocks toppled, crushing houses. Roofs were torn off and tossed into the woods. And farther to the east, the devastation was astounding. When my kids started to complain about the minor hardships they had to endure—going without their Xbox for a few days until power was restored—my wife and I drove them down to Staten Island so they could see what real hardship was like, and so, young as they were, they could pitch in with us and help in whatever small way we could with the cleanup.

I've revisited that moment on the deck more times than I care to count in the two years since the storm. In a way, it's become a kind of metaphor for me, not of an idea but of a question. My home, my family, my stuff, all of it was spared from damage or destruction by a fluke of topography and an accident of design. Nothing I ever intentionally did put me in better stead to escape the storm than my neighbors or my friends in places like Hoboken or on the New Jersey shore. On the contrary, it could be argued that some of my choices—moving to a comparatively remote woodland where services are strained even under normal circumstances, for example—should have made me more vulnerable.

In fact, if you heed the scientists who warn that our consumption of fossil fuels is contributing to the wildly unstable climate—and I do—then you could argue that by living in a place where my consumption is by necessity greater than that of my friends in more urban areas, my family and I might be in some small way more responsible for the devastation that we escaped.

But all the same, we escaped it. This time. And that raises the question: What is my responsibility to understand and address to whatever extent I can the chain of events that perhaps contributed to the savagery of the storm, and certainly contributed to the brutality of its impact?

In a way, my deck, the 200-square-foot redwood island of calm in the storm, was a lot like most of the United States at the moment. We have, as a nation, made choices that by all reasonable expectations should have put us in harm's way. There is little doubt that we continue to make choices that are likely to make the danger even greater. And yet, by dint of an accident of geography and economics, we have so far been spared the worst consequences of our actions. And even as those consequences begin to take hold in other places, here, in the parts of America where most of us live, at least for the moment, we can hear the winds roaring over our heads like that coal train, but somehow the worst of the danger still seems removed. What is our responsibility?

If there was any question I wanted to explore in this book, it was that. Not as a scientist. I'm not one. Not as a theorist. I'm not one of those either. I'm just a storyteller.

Whatever failings there are in this book, and I'm sure there are quite a few, are entirely my own. Whatever strengths there might be have little to do with me. For that reason, there are a lot of people I need to thank for helping me tell the stories in this book.

First of all, there's my agent, Byrd Leavell. Once Sandy had slammed into New York and focused his attention, he did an incredible job of honing my questions and sharpening my thoughts, and he pushed me mightily to turn a vague notion into a book.

Even still, that notion would have gone nowhere if it hadn't been for the diligence and commitment of the folks at the University of Texas Press, Gianna LaMorte and my editor, Casey Kittrell, in particular. And I owe a special debt of gratitude to the anonymous readers who reviewed the manuscript in its earliest form and whose comments helped me to make it better.

Early on, I made a decision to write this book in public, and so, time and again, as I was sequestered in my fetid office, scribbling away, I posted experimental passages on Facebook and elsewhere. The feedback I got from my many friends—some of them longtime real-world friends, some of them recent acquaintances, others virtual friends—was invaluable. Thank you, Joe Henderson and Don Duggan-Hass, Doug Heuck, Bruce Selleck, Marc Ainger, Tom Frost, Todd Tanner, Jesse Waters, Heather Brambilla, Simona Perry, Dennis McGrath, John LaRose, Dave Petrenka, Corky Siezmaszko—the list goes on for days.

My loving wife often points out, as gently as possible, that whenever I am working on a book I become a thundering bastard. She's right. And I am eternally grateful to her and to my kids for putting up with that.

I also need to express my gratitude to Professor Terry Engelder for making some important introductions. And it goes without saying that I am utterly indebted to everyone who makes an appearance here. This book is a collection of their stories, not mine. They've all become heroes of mine.

My old man,
James Joseph McGraw.

Finally, I want to specifically thank a guy who appears several times in the book but isn't around to hear me thank him. He's the guy who taught me how to tell a story, and more importantly, he's the guy who taught me how to listen to one. He's the guy who taught me how to listen to the wind.

In the end, this book is for him. My dad. James Joseph McGraw, July 17, 1932–December 28, 1998.

Thanks, Dad.

Notes

Chapter 1

1. John Cook et al., "Quantifying the Consensus on Anthropogenic Global Warming in the Scientific Literature," *Environmental Research Letters* 8, no. 2 (2013), doi:10.1088/1748-9326/8/2/024024[0].

2. U.S. Energy Information Administration, *Monthly Energy Review*, April 5, 2013.

Chapter 2

1. Andrew Freedman, "Hurricane Irene Ranked Most Costly Category One Storm," Climatecentral.org, May 10, 2012.

2. Sean A. Marcott et al., "A Reconstruction of Regional and Global Temperature for the past 11,300 Years," *Science*, March 8, 2013.

3. Mathew Barrett Gross and Mel Gilles, "How Apocalyptic Thinking Prevents Us from Taking Political Action," *Atlantic*, April 23, 2012.

4. Yale Project on Climate Change Communication, *Global Warming's Six Americas in September 2012*.

Chapter 3

1. David C. Barker and David H. Bearce, "End-Times Theology, the Shadow of the Future, and Public Resistance to Addressing Global Climate Change," *Political Research Quarterly*, May 1, 2012, doi:10.1177/1065912912442243.

Chapter 5

1. William M. Welch, Gary Strauss (*USA Today*), and the *Arizona Republic* staff, "19 Firefighters Killed in Arizona Fire, Now at 8,400 Acres," *USA Today*, July 1, 2013.

2. Jim Melewitz, "West's Record Wildfires Raise Questions about Development," *USA Today*, August 19, 2013.

3. Chesapeake Beach Consulting, *2012 National Survey of Hunters and Anglers* (prepared for the National Wildlife Federation), September 25, 2012.

Chapter 6

1. Executive Office of the President, "The President's Climate Action Plan," June 2013.

2. Albert Gore, "Obama's Climate Speech: 'It Is Time for Congress to Share His Ambition,'" *Guardian*, June 25, 2013.

3. Theda Skocpol, "Naming the Problem: What It Will Take to Counter Extremism and Engage Americans in the Fight against Global Warming" (paper prepared for "The Politics of America's Fight against Global Warming" symposium cosponsored by the Columbia School of Journalism and the Scholars Strategy Network, February 14, 2013, Harvard University).

4. Theda Skocpol, "Learning from the Cap and Trade Debate," *Grist*, March 6, 2013.

Chapter 7

1. Intergovernmental Panel on Climate Change (IPCC), *Fifth Assessment Report* (Geneva, Switzerland: IPCC, 2013).

2. Stefan Rahmstorf, Grant Foster, and Anny Cazenave, "Comparing Climate Projections to Observations up to 2011," *Environmental Research Letters* 7, no. 4 (2012).

Chapter 8

1. NYC Special Initiative for Rebuilding and Resiliency, *A Stronger, More Resilient New York* (New York: Office of the Mayor, 2013).

2. Michael Mann, *The Hockey Stick and the Climate Wars: Dispatches from the Front Lines* (New York: Columbia University Press, 2012).

3. NOAA Drought Task Force Narrative Team, *An Interpretation of the Origins of the 2012 Central Great Plains Drought Assessment Report* (March 20, 2013).

Chapter 9

1. A. G. Sulzberger, "Army Corps Blows Up Missouri Levee," *New York Times*, May 2, 2011.

2. Stephan Saunders et al., *Doubled Trouble: More Midwestern Extreme Storms* (published by the Rocky Mountain Climate Organization and the Natural Resources Defense Council, May 16, 2012).

3. Albert Gore, *Earth in the Balance* (Boston: Houghton-Mifflin, 1992).

4. Albert Gore, *The Future: Six Drivers of Global Change* (New York: Random House, 2013).

Chapter 10

1. Philip Shabecoff, "Global Warming Has Begun, Expert Tells Senate," *New York Times*, June 24, 1988.

2. Stephan Stromberg, "Another Climate Change Warning, Written in the Shells of Crabs," *Washington Post*, April 8, 2013.

3. National Hurricane Center, "Tropical Cyclone Report, Hurricane Sandy (AL182012)," February 12, 2013.

4. Andrew C. Kemp and Benjamin Horton, "Contribution of Relative Sea Level Rise to Historical Hurricane Flooding in New York City," *Journal of Quaternary Science* 28, no. 6 (August 2013).

Chapter 11

1. *2007 Census of Agriculture* (prepared and released in 2009 by the USDA).

2. Erik O'Donoghue et al., *The Changing Organization of U.S. Farming*, USDA Economic Information Bulletin no. 88 (Washington, DC: U.S. Department of Agriculture, Economic Research Service, December 2011).

Chapter 12

1. Richard B. Alley, *Earth: The Operators' Manual* (New York: W. W. Norton, 2011).

2. World Meteorological Organization, *The Global Climate 2001–2010: A Decade of Climate Extremes* (Geneva, Switzerland: WMO, 2013).

3. Government Accountability Office, *GAO 2013 High-Risk Series: An Update*, GAO-13-359T (February 14, 2013).

4. Michael Levi, *The Power Surge: Energy, Opportunity, and the Battle for America's Future* (New York: Oxford University Press, 2013).

5. Bill McKibben, "Global Warming's Terrifying New Math," *Rolling Stone*, July 19, 2012.

6. Adam R. Brandt, Adam Millard-Ball, Matthew Ganser, and Steven M. Gorelick, "Peak Oil Demand: The Role of Fuel Efficiency and Alternative Fuels in a Global Oil Production Decline," *Environmental Science and Technology Letters* 47, no. 14 (2013): 8031-8041.

7. Andrew Revkin, "More Signs of 'Peak Us' in New Study of 'Peak Oil Demand,'" *NY Times Dot Earth* (blog), July 10, 2013.

8. White House Council of Economic Advisors, "Climate Change and the Path toward Sustainable Energy Sources," chapter 6 of *2013 Economic Report of the President* (March 2013).

Bibliography

Chapter 1

Cook, John, Dana Nuccitelli, Sarah A Green, Mark Richardson, Bärbel Winkler, Rob Painting, Robert Way, Peter Jacobs, and Andrew Skuce. "Quantifying the Consensus on Anthropogenic Global Warming in the Scientific Literature." *Environmental Research Letters* 8, no. 2 (2013), doi:10.1088/1748-9326/8/2/024024.

U.S. Energy Information Administration. *Monthly Energy Review*, April 5, 2013.

Chapter 2

Freedman, Andrew. "Hurricane Irene Ranked Most Costly Category One Storm." Climatecentral.org, May 10, 2012.

Gross, Mathew Barrett, and Mel Gilles. "How Apocalyptic Thinking Prevents Us from Taking Political Action." *Atlantic*, April 23, 2012.

Marcott, Sean A., Jeremy D. Shakun, Peter U. Cark, and Alan C. Mix. "A Reconstruction of Regional and Global Temperature for the Past 11,300 Years." *Science*, March 8, 2013.

NCADAC. *Climate Change Impacts in the United States.* Third National Climate Assessment. January 2013.

Yale Project on Climate Change Communication. *Global Warming's Six Americas in September 2012.*

Chapter 3

Barker, David C., and David H. Bearce. "End-Times Theology, the Shadow of the Future, and Public Resistance to Addressing Global Climate Change." *Political Research Quarterly*, May 1, 2012, doi:10.1177/1065912912442243.

Gross, Mathew Barrett, and Mel Gilles. *The Last Myth: What the Rise of Apocalyptic Thinking Tells Us about America*. Amherst, NY: Prometheus Books, 2012.

Landes, Richard. *Heaven on Earth: The Varieties of the Millennial Experience*. New York: Oxford University Press, 2011.

Chapter 4

Hayhoe, Katharine, and Andrew Farley. *A Climate for Change: Global Warming Facts for Faith-Based Decisions*. New York: FaithWords/Hachette Book Group, 2009.

Intergovernmental Panel on Climate Change (IPCC). *Climate Change 2007: Synthesis Report*. Geneva, Switzerland: IPCC, 2007.

Chapter 5

Chesapeake Beach Consulting. *2012 National Survey of Hunters and Anglers*. Prepared for the National Wildlife Federation, September 25, 2012.

Melewitz, Jim. "West's Record Wildfires Raise Questions about Development." *USA Today*, August 19, 2013.

Welch, William M., Gary Strauss (*USA Today*), and the *Arizona Republic* staff. "19 Firefighters Killed in Arizona Fire, Now at 8,400 Acres." *USA Today*, July 1, 2013.

Chapter 6

Executive Office of the President. "The President's Climate Action Plan." June 2013.

Gore, Albert. "Obama's Climate Speech: 'It Is Time for Congress to Share His Ambition.'" *Guardian*, June 25, 2013.

Skocpol, Theda. "Learning from the Cap and Trade Debate." *Grist*, March 6, 2013.

———. "Naming the Problem: What It Will Take to Counter Extremism and Engage Americans in the Fight against Global Warming." Paper prepared for "The Politics of America's Fight against Global Warming" symposium cosponsored by the Columbia School of Journalism and the Scholars Strategy Network, February 14, 2013, Harvard University.

Chapter 7

Intergovernmental Panel on Climate Change (IPCC). *Fifth Assessment Report*. Geneva, Switzerland: IPCC, 2013.

Rahmstorf, Stefan, Grant Foster, and Anny Cazenave. "Comparing Climate Projections to Observations up to 2011." *Environmental Research Letters* 7, no. 4 (2012).

Chapter 8

Mann, Michael. *The Hockey Stick and the Climate Wars: Dispatches from the Front Lines*. New York: Columbia University Press, 2012.

NOAA Drought Task Force Narrative Team. *An Interpretation of the Origins of the 2012 Central Great Plains Drought Assessment Report*, March 20, 2013.

NYC Special Initiative for Rebuilding and Resiliency. *A Stronger, More Resilient New York*. New York: Office of the Mayor, 2013.

Chapter 9

Gore, Albert. *Earth in the Balance*. Boston: Houghton-Mifflin, 1992.

———. *The Future: Six Drivers of Global Change*. New York: Random House, 2013.

Saunders, Stephan, Dan Finlay, and Tom Easley (principal authors) and Theo Spencer (contributing author). *Doubled Trouble: More Midwestern Extreme Storms*. Published by the Rocky Mountain Climate Organization and the Natural Resources Defense Council, May 16, 2012.

Sulzberger, A. G. "Army Corps Blows Up Missouri Levee." *New York Times*, May 2, 2011.

Chapter 10

Kemp, Andrew C., and Benjamin Horton. "Contribution of Relative Sea Level Rise to Historical Hurricane Flooding in New York City." *Journal of Quaternary Science* 28, no. 6 (August 2013).

National Hurricane Center. "Tropical Cyclone Report, Hurricane Sandy (AL182012)." February 12, 2013.

Shabecoff, Philip. "Global Warming Has Begun, Expert Tells Senate." *New York Times*, June 24, 1988.

Stromberg, Stephan. "Another Climate Change Warning, Written in the Shells of Crabs." *Washington Post*, April 8, 2013.

Chapter 11

O'Donoghue, Erik J., Robert A. Hoppe, David E. Banker, Robert Ebel, Keith Fuglie, Penni Korb, Michael Livingston, Cynthia Nickerson, and Carmen Sandretto. *The Changing Organization of U.S. Farming*. USDA Economic Information Bulletin no. 88. Washington, D.C.: U.S. Department of Agriculture, Economic Research Service, December 2011.

2007 Census of Agriculture. Prepared and released in 2009 by the U.S. Department of Agriculture.

Chapter 12

Alley, Richard B. *Earth: The Operators' Manual*. New York: W. W. Norton, 2011.

Brandt, Adam R., Adam Millard-Ball, Matthew Ganser, and Steven M. Gorelick. "Peak Oil Demand: The Role of Fuel Efficiency and Alternative Fuels in a Global Oil Production Decline." *Environmental Science and Technology* 47, no. 14 (2013): 8031–8041.

Government Accountability Office. *GAO 2013 High-Risk Series: An Update*. GAO-13-359T. February 14, 2013.

Levi, Michael. *The Power Surge: Energy, Opportunity, and the Battle for America's Future*. New York: Oxford University Press, 2013.

McKibben, Bill. "Global Warming's Terrifying New Math." *Rolling Stone*, July 19, 2012.

Revkin, Andrew. "More Signs of 'Peak Us' in New Study of 'Peak Oil Demand.'" *NY Times Dot Earth* (blog), July 10, 2013.

White House Council of Economic Advisors. "Climate Change and the Path

toward Sustainable Energy Sources." Chapter 6 of *2013 Economic Report of the President*. Delivered to Congress March 2013.
World Meteorological Organization. *The Global Climate 2001–2010: A Decade of Climate Extremes*. Geneva, Switzerland: WMO, 2013.

Index

Page numbers in italics indicate photographs.